第十五届建筑物理学术大会论文集
双碳战略背景下的建筑物理

中国建筑学会建筑物理分会
同济大学建筑与城市规划学院
中国建筑科学研究院有限公司
上海市建筑科学研究院有限公司　　主编
亚热带建筑与城市科学全国重点实验室
绿色建筑全国重点实验室
高密度人居环境生态与节能教育部重点实验室

同济大学出版社
TONGJI UNIVERSITY PRESS
· 上海 ·

图书在版编目（C I P）数据

第十五届建筑物理学术大会论文集：双碳战略背景下的建筑物理 / 中国建筑学会建筑物理分会等主编 . -- 上海：同济大学出版社，2024.4

ISBN 978-7-5765-1000-3

I. ①第… II. ①中… III. ①建筑学－物理学－学术会议－文集 IV. ① TU11-53

中国国家版本馆 CIP 数据核字 (2024) 第 072996 号

第十五届建筑物理学术大会论文集

双碳战略背景下的建筑物理

中国建筑学会建筑物理分会
同济大学建筑与城市规划学院
中国建筑科学研究院有限公司
上海市建筑科学研究院有限公司　　主编
亚热带建筑与城市科学全国重点实验室
绿色建筑全国重点实验室
高密度人居环境生态与节能教育部重点实验室

责任编辑 由爱华　朱笑黎　**责任校对** 徐春莲　**封面设计** 张　微

出版发行　同济大学出版社　www.tongjipress.com.cn
（地址：上海市四平路 1239 号　邮编：200092　电话：021-65985622）
经　　销　全国各地新华书店
印　　刷　苏州市古得堡数码印刷有限公司
开　　本　787 mm×1092 mm　1/16
印　　张　12.25
字　　数　338 000（纸书）　2 720 000（线上资源）
版　　次　2024 年 4 月第 1 版
印　　次　2024 年 4 月第 1 次印刷
书　　号　ISBN 978-7-5765-1000-3
定　　价　68.00 元

前　言

为积极响应和深入贯彻党的二十大提出的“打造宜居、韧性、智慧城市”的战略部署，进一步推动城镇化领域的科技创新发展，建设集约、智能、绿色、低碳的建筑和城市，“第十五届建筑物理学术大会”于2023年11月17至19日在上海举办，大会主题为“双碳战略背景下的建筑物理”。

本届会议得到国内相关高校、科研院所、企事业单位和广大科技工作者的热情支持与积极参与，共950余人参会，收到论文投稿463篇，经专家评阅，录用316篇，其中：建筑热工与节能71篇，建筑与环境声学42篇，建筑采光与照明54篇，绿色建筑环境品质28篇，建筑门窗与幕墙18篇，城市微气候46篇，健康智慧的城市与建筑23篇，低碳城市与建筑34篇。这些论文聚焦建筑物理领域学术研究和科技发展热点，探讨了面临的机遇与挑战。

本论文集的编辑出版工作是在中国建筑学会建筑物理分会的指导下进行的。建筑物理领域的许多专家学者为论文的评审和出版工作做出了重要贡献，保证了论文的质量。同济大学出版社的编辑们为论文集的出版付出了辛勤的劳动。在此感谢各位专家学者的无私奉献，并对给予本届会议大力支持、热情帮助并积极参与的社会各界人士表示由衷的感谢！

编委会水平有限，对于论文集可能存在的错误，欢迎各界人士批评指正。

本书编委会

2024年3月

目　录

第 1 篇　　建筑热工与节能

第 2 篇　建筑与环境声学

第 4 篇　绿色建筑环境品质

第 5 篇　建筑门窗与幕墙

第 6 篇 城市微气候

第 8 篇　低碳城市与建筑

第 1 篇

建筑热工与节能

方案初期办公建筑生形方法研究

Study on the Forming Method of Office Building at the Initial Stage of the Scheme

廉慧敏，李紫微 *
Lian Huimin, Li Ziwei

摘要 方案初期的节能设计能够有效提升方案的节能特性，然而形体单一是制约方案初期设计生成的重要原因之一。本研究搭建了基于性能模拟的优化流程，形成方案初期办公建筑的形体生成逻辑，并通过遗传算法实现形体生成。为缩短计算时间，本文采用 ISO 13790 中的准稳态计算方法进行能耗公式的程序编写，并对方案初期的建筑形体进行能耗计算；在此基础上运用遗传算法寻找方案初期能耗最低的建筑形体。研究结果表明，建筑能耗受中庭形态和立面凹凸的影响，当建筑没有中庭且立面凹凸越少时，建筑能耗越低。

关键词 方案初期；生形逻辑；形体优化；办公建筑

Abstract The energy-saving design in the initial stage of the scheme can effectively improve the energy-saving characteristics of the scheme, but the single form is one of the important factors restricting the generation of the initial design of the scheme. In this study, an optimization process based on performance simulation was established to form the form generation logic of the initial office building and the form generation was realized by genetic algorithm. In order to shorten the calculation time, the quasi-steady state calculation method in ISO 13790 is used to program the energy consumption formula and calculate the energy consumption of the initial building form. On this basis, a genetic algorithm is used to find the building form with the lowest energy consumption in the initial stage of the scheme. The results show that the building energy consumption is affected by the shape of the atrium and the convex of the facade. The building energy consumption is lower when the building has no atrium and the fewer convex of the facades.

Keywords Initial stage of the program; Form logic; Shape optimization; Office building

廉慧敏，北京工业大学，硕士研究生。
李紫微，北京工业大学，城市建设学部研究生导师，副教授。

极端天气事件对建筑能耗的影响

Impact of Extreme Climate on Building Energy Consumption

韩雨菲，刘大龙，刘紫琪，冯本成，孟智豪
Han Yufei, Liu Dalong, Liu Ziqi, Feng Bencheng, Meng Zhihao

摘要 本文研究了极端天气对建筑能耗的影响，发现干球温度和直射辐射是关键气候参数。在极端冷气象下，昆明的采暖能耗增长率最大，而哈尔滨的采暖能耗变化率最大且在特定时间段内变化速度最快。在极端热气象下，广州的空调能耗增长率和变化率最大，且在特定时间段内空调能耗变化速度最快。总体而言，极端气候对建筑能耗的影响具有显著的地域特性，不同地区能耗的增长程度和变化速度也不同。由极端冷气象导致的建筑能耗增加在严寒地区和寒冷地区更为明显，而由极端热气象导致的能耗增加在夏热冬暖和夏热冬冷地区更为明显。

关键词 极端天气；建筑能耗；温度；太阳辐射

Abstract This paper investigates the effect of extreme weather on building energy consumption and finds that dry bulb temperature and direct radiation are key climate parameters. Under extremely cold weather, the growth rate of heating energy consumption in Kunming is the largest, while the rate of change of heating energy consumption in Harbin is the largest and the fastest within a specific period. In extremely hot weather, Guangzhou has the largest growth rate and rate of change in air conditioning energy consumption, and the fastest rate of change in air conditioning energy consumption within a specific period. Overall, the impact of extreme weather on building energy consumption has significant geographical characteristics, and the degree of growth and rate of change of energy consumption varies from region to region. The increase in building energy consumption caused by extremely cold weather is more pronounced in severely cold regions and cold regions, while the increase in energy consumption caused by extremely hot weather is more pronounced in hot summer and warm winter and hot summer and cold winter regions.

Keywords Extreme weather; Building energy consumption; Temperature; Solar radiation

韩雨菲，女，西安建筑科技大学，硕士研究生。
刘大龙，男，西安建筑科技大学，博士，副教授，研究生导师。
刘紫琪，女，西安建筑科技大学，硕士研究生。
冯本成，西安建筑科技大学，硕士研究生。
孟智豪，西安建筑科技大学，硕士研究生。

辐射制冷涂料与加气混凝土的降温隔热效果对比测试

Experimental Comparison of the Cooling/Insulation Effect between Radiative Cooling Coating and Aerated Concrete

雷玥，何悦，李小龙，郑星，冯驰 *
Lei Yue, He Yue, Li Xiaolong, Zheng Xing, Feng Chi

雷玥，重庆大学建筑城规学院，博士研究生。
何悦，重庆大学建筑城规学院，博士研究生。
李小龙，重庆大学建筑城规学院，助理研究员。
郑星，苏黎世联邦理工学院新加坡中心，助理研究员。
冯驰，重庆大学建筑城规学院，教授。

摘要 辐射制冷涂料能有效降低建筑围护结构的外表面温度。为探究辐射制冷涂料在实际应用中的降温效果，本研究分别搭建了覆涂辐射制冷涂料和附着加气混凝土层的金属夹芯板屋面模型，并以裸露的金属夹芯板屋面模型为对照组，在重庆夏季极高温天气下进行现场测试。结果表明，全天覆涂辐射制冷涂料的屋面模型内表面平均温度低于附着加气混凝土层和裸露屋面模型的内表面温度 0.7℃和 0.8℃，外表面温度最多可低于环境温度 7.1℃；日间平均降温幅度达 2.1℃。在正午时段内，附着加气混凝土层的屋面模型外表面温度为三者中最高，最高可超过环境温度 26.7℃。

关键词 辐射制冷涂料；加气混凝土；降温效果；被动节能

Abstract The radiative cooling coating can effectively reduce the exterior surface temperature of the building envelope. To investigate the cooling effect of radiative cooling coating in actual practice, this paper built two models of metal sandwich panel roofs decorated with radiative cooling coating and aerated concrete respectively, with a bare one as the control. A field measurement was carried out during the extremely hot summer in Chongqing. The results show that the daily average interior surface temperature of the roof with radiative cooling coating was 0.7 °C and 0.8 °C lower than those of the roof with aerated concrete and the bare roof, respectively, and the exterior surface of the roof with radiative cooling coating had a maximum subambient temperature of 7.1 °C. The daytime average cooling temperature was 2.1 °C. During the noontime, the exterior surface temperature of the roof with aerated concrete was the highest, exceeding the ambient air temperature by up to 26.7 °C.

Keywords Radiative cooling coating; Aerated concrete; Cooling effect; Passive energy saving

---

严寒地区零能耗建筑保温结构一体化墙板热桥研究

Research on the Thermal Bridge of Wall Panel of Integration of Thermal Insulation and Building Structure of Zero-Energy Building in Severe Cold Areas

张翼，邓琴琴，孙立新，邹瑜
Zhang Yi, Deng Qinqin, Sun Lixin, Zou Yu

张翼，中国建筑科学研究院有限公司，硕士研究生。
邓琴琴，中国建筑科学研究院有限公司，研究员。
孙立新，中国建筑科学研究院有限公司，研究员。
邹瑜，中国建筑科学研究院有限公司，教授级高工。

摘要 严寒地区寒冷干燥，而热桥部位具有较高的传热系数，会导致局部区域的热流密度增大，使室内温度不均衡，造成能量损失，故减少热桥的影响将大幅提升围护结构的热工性能。本文以严寒地区某零能耗保温结构一体化建筑为研究对象，通过建筑能耗模拟及热桥节点分析，研究一体化墙板和建筑热桥节点对能耗的影响。通过定性和定量的比对分析来观察热桥对建筑能耗影响的变化情况，进一步探讨合理的局部保温等方法来消除热桥产生的影响，同时探究了外设光伏板形成的空腔对外墙温度场的影响，并给出相关建议。

关键词 严寒地区；零能耗建筑；保温结构一体化；热桥

Abstract The climate in the severe cold areas is cold and dry, and the thermal bridge has a high heat transfer coefficient, which will lead to the increase of heat flux in the local area of the building, make the indoor temperature imbalance, and cause energy loss. Thus, reducing the influence of thermal bridges will greatly improve the thermal performance of the envelope structure. This paper takes a zero-energy building with integrated thermal insulation and building structure in severe cold regions as the research object. Through building energy simulation and thermal bridge node analysis, the influence of integrated wall panels and thermal bridge nodes on energy consumption is studied. Through qualitative and quantitative comparative analysis, the change of the influence of thermal bridges on building energy consumption is observed, and reasonable local insulation and other methods are further discussed to eliminate the influence of thermal bridges. At the same time, the influence of the cavity formed by the peripheral photovoltaic panel on the external wall temperature field is explored and relevant suggestions are given.

Keywords Severe cold areas; Zero-energy building; Integration of thermal insulation and building structure; Thermal bridge

长沙地区相变特朗勃墙结构的节能效果及热性能分析

Energy Saving Effect and Thermal Performance of a Trombe Wall with Phase Change Materials in Changsha

张圆，孙小琴，朱江恬，廖柏琛

Zhang Yuan, Sun Xiaoqin, Zhu Jiangtian, Liao Baichen

摘要 本文对长沙某轻质建筑进行相变特朗勃墙改造和热性能对比实验分析，运用 EnergyPlus 软件对传统建筑、特朗勃墙建筑、相变特朗勃墙建筑进行全年能耗模拟，对比分析不同相变温度的相变材料、不同安装方式和不同通风方式下内壁面温度和特朗勃墙进出风口温度变化。结果显示：在夏热冬冷地区，特朗勃墙结合内侧布置的 PCM22 并合理开启通风具有最好的节能效果。相对传统建筑能耗的全年能耗、供暖能耗以及制冷能耗的节能率分别为 42.09%、22.47% 和 31.13%。

关键词 相变材料；特朗勃墙；节能效果；内壁面温度；进出风口温度

Abstract A light building in Changsha was transformed with a Trombe wall with phase change materials (PCMs) and the thermal performance was analyzed in this paper. EnergyPlus was used to simulate the annual energy consumption of traditional buildings, Trombe wall buildings and buildings with Trombe walls combined with PCMs. The changes of inner wall surface temperature and Trombe wall inlet and outlet air temperature under different phase change temperatures, different installation modes, and different ventilation modes were compared and analyzed. The results show that in hot summer and cold winter regions, the Trombe wall combined with PCM22 on the inside and ventilation possesses the best energy-saving effect. Compared to the traditional building energy consumption, the annual, heating and cooling energy saving rates of the building with Trombe wall combined with PCMs are 42.09%, 22.47% and 31.13%, respectively.

Keywords Phase change materials; Trombe wall; Energy saving effect; Indoor air temperature; Inlet and outlet air temperature

张圆，长沙理工大学，硕士研究生。
孙小琴，长沙理工大学能源与动力工程学院，教授，副院长。
朱江恬，长沙理工大学，硕士研究生。
廖柏琛，长沙理工大学，硕士研究生。

建筑绝热材料特征热物性参数测试值尺寸效应研究

Size Effect of Characteristic Thermophysical Parameters Test Values of Building Insulation Materials

杨雯*，张冠杰，史子涵

Yang Wen, Zhang Guanjie, Shi Zihan

摘要 试件尺寸是实验获取建筑绝热材料特征热物性参数的关键影响因素之一。本文选取两种代表性传统建筑绝热材料，通过实验探究热湿耦合作用下试件三维尺寸与基本热物性参数、导出热物性参数的变随动关系。结果表明，低湿条件下试件厚度对测试结果的影响程度较大，高湿条件下试件底面积对测试结果的影响较大，EPS 在中低湿条件下测定导热系数建议尺寸为 50 mm×50 mm×30 mm。在不同相对湿度工况条件下比热容与蓄热系数的尺寸效应变化趋势相同，但热扩散系数的变化趋势完全相反。

关键词 建筑绝热材料；热物性参数；试件尺寸；相对湿度

Abstract The size of the specimen is one of the key influencing factors for obtaining the characteristic thermophysical parameters of building insulation materials through experiments. This article selects two representative traditional building insulation materials and explores the variation relationship between the three-dimensional size of the specimen and the basic thermophysical parameters under the coupling effect of heat and moisture through experiments, as well as the derived thermophysical parameters. The results indicate that the thickness of the specimen has a significant impact on the test results under low humidity conditions, and the bottom area of the specimen has a significant impact on the test results under high humidity conditions. The recommended size for measuring the thermal conductivity of EPS under medium to low humidity conditions is 50mm×50mm×30mm. The changing trend of the size effect of specific heat capacity and heat storage coefficient under different relative humidity conditions is the same, but it is opposite to the changing trend of the thermal diffusion coefficient.

Keywords Building insulation materials; Thermophysical parameters; Specimen size; Relative humidity

杨雯，西安建筑科技大学，副教授。
张冠杰，西安建筑科技大学，硕士研究生。
史子涵，西安建筑科技大学，硕士研究生。

建筑表皮中热致变色材料的变色特性与节能效益研究

Study on the Discoloration Characteristics and Energy-Saving Benefits of Thermochromic Materials in the Building Envelope

胡晨曦，刘书含，孟曦
Hu Chenxi, Liu Shuhan, Meng Xi

胡晨曦，青岛理工大学滨海人居环境学术创新中心，硕士研究生。
刘书含，日本北九州市立大学环境工学院，博士研究生。
孟曦，青岛理工大学滨海人居环境学术创新中心，教授、博士生导师。

摘要 传统建筑表皮材料无法满足热需求季节性差异的要求，热致变色材料通过受热变色可改变太阳辐射的吸收程度，在一定程度上满足冷热负荷需求的差异性。本研究测试了不同颜色材料的光学特性并通过数值模拟的方式研究了不同季节热致变色材料对建筑热环境的影响机理。研究表明，热致变色材料在建筑中的适应性较好且有较强节能潜力，但不同颜色热致变色材料的变色特性差异较大。

关键词 热致变色材料；季节适应性；热性能；颜色

Abstract Traditional building skin materials cannot meet the requirements of seasonal differences in thermal demand. Thermochromic materials (TCM) can change the degree of solar radiation absorption by changing their color in response to heat, thus satisfying the seasonal differences. In this study, the optical properties of different colored materials were tested and the effect of TCMs on the thermal environment of buildings in different seasons was investigated by means of numerical simulations; the study shows that TCMs are well adapted to buildings and have strong energy saving potential, but the color change characteristics of different colored TCMs vary greatly.

Keywords Thermochromic materials (TCM); Seasonal adaptability; Thermal performance; Colors

---

全要素气候模拟实验舱在风驱雨湿热过程研究中的应用

Application of All-Element Climate Simulation Experimental Cabin in the Study of Wind-Driven Rain, Humidity and Heat Processes

李丽，李紫英，邹子琪，李元奎，林卫新
Li Li, Li Ziying, Zou Ziqi, Li Yuankui, Lin Weixin

李丽，广州大学建筑与城市规划学院，副教授。
李紫英，广州大学建筑与城市规划学院，硕士研究生。
邹子琪，广东省建设工程质量安全检测总站有限公司，助理工程师。
李元奎，广州大学建筑与城市规划学院，实验员。
林卫新，广州大学建筑与城市规划学院，实验员，实验室主任。

摘要 为研究风驱雨对建筑围护结构湿热过程的影响，文章总结分析了运用实验法取得的研究成果以及相关实验平台的搭建，在此基础上，搭建可模拟光照、风、雨等自然界综合气候环境条件的全要素气候模拟实验舱，对其进行热湿传递稳态和动态实验模拟测试，并进行实验应用。测试结果表明，该模拟实验舱稳定性好，模拟精度高，可实现对室外气象环境复现。

关键词 全要素气候模拟实验舱；湿热过程；风驱雨；建筑围护结构

Abstract In order to study the effect of wind-driven rain on the hygrothermal process of the building envelope, the article summarises and analyses the research results obtained by using the experimental method and the construction of the relevant experimental platform. On this basis, an all-factor climate simulation chamber, which can simulate light, wind and rain and other natural comprehensive climate conditions, is built and tested for steady-state and dynamic experimental simulation of heat and humidity transfer, and then experimentally applied. The test results show that the simulation chamber has good stability and high simulation accuracy, and can achieve the reproduction of outdoor weather environment.

Keywords All-factor climate simulation experimental cabin; Moist heat process; Wind-Driving Rain; Building envelope

光伏相变复合围护结构发电及隔热性能分析

Analysis of Power Generation and Heat Insulation Performance of Composite Building Envelope Integrated Photovoltaic Modules and Phase Change Materials

蔡阳，刘子瑞，洪炳华，董明浩，赵福云
Cai Yang, Liu Zirui, Hong Binghua, Dong Minghao, Zhao Fuyun

蔡阳，暨南大学，副教授。
刘子瑞，董明浩，暨南大学国际能源学院，本科生。
洪炳华，暨南大学国际能源学院，硕士研究生。
赵福云，武汉大学动力与机械学院，教授。

摘要 光伏相变复合围护结构因兼具光伏发电与建筑相变蓄热调节作用，受到广泛关注。基于光伏发电和相变蓄热耦合特性，本研究构建了光伏相变复合围护结构的物理模型，探究太阳辐射强度和相变围护结构厚度等参数对光伏复合围护结构发电及隔热性能的影响。通过比较传统结构和复合围护结构之间的性能差异，阐述该结构发电及隔热综合性能。研究表明，该围护结构的综合性能与太阳辐射强度、相变厚度紧密相关。相比于传统围护结构，复合围护结构性能显著提升。

关键词 光伏相变复合围护结构；发电性能；隔热性能；建筑节能

Abstract Building envelope integrated photovoltaic modules and phase change materials has been widely concerned because of its functions of both photovoltaic power generation and building phase change thermal storage. Based on the coupled characteristics of photovoltaic power generation and phase change heat storage, a physical model of the proposed building envelope is presented, and the effects of irradiation intensity and phase change material thickness on the power generation and heat insulation performance are further investigated. By comparing the performance differences between the traditional structure and the composite envelope structure, the comprehensive performance of this composite building envelope is revealed through power generation efficiency and internal wall temperature. The results show that the performance of the presented envelope is closely relevant to the irradiation intensity and phase change material thickness. Compared with traditional enclosure structures, the performance of composite building envelopes is significantly improved for the current study.

Keywords Building envelope integrated photovoltaic modules and phase change materials; Power generation performance; Heat Insulation performance; Building energy conservation

---

“双碳”目标下外墙保温构造体系应用现状及发展趋势

The Application and Trend of Thermal Insulation Systems for External Wall under the Goals of Carbon Peaking and Carbon Neutrality

孙立新，杨玉忠，张昭瑞，赵矗，杨寒羽
Sun Lixin, Yang Yuzhong, Zhang Zhaorui, Zhao Chu, Yang Hanyu

孙立新，中国建筑科学研究院有限公司，正高级工程师，碳中和研究院副总工程师。
杨玉忠，建科环能科技有限公司，研究员，建筑节能测评中心主任。
张昭瑞，建科环能科技有限公司，高级工程师。
赵矗，建筑环境与安全国家重点实验室，高级工程师。
杨寒羽，建筑环境与安全国家重点实验室，助理工程师。

摘要 本文依据超低能耗建筑和零碳建筑等高性能建筑对外墙保温构造体系的新发展需求，结合全国外墙保温工程质量和行业政策调研，分析了我国外墙保温构造技术标准体系的工程与材料应用现状、常见工程质量问题。基于“双碳”目标，分析了薄抹灰外保温体系、保温装饰一体化体系、保温结构一体化体系等构造应用场景及潜在问题，提出了技术和政策性建议。

关键词 “双碳”目标；外墙保温构造体系；应用现状；工程质量问题；趋势；建议

Abstract This paper analyzes the engineering and material application status and common engineering quality problems of thermal insulation systems for external walls through the requirements research and the nationwide survey of project quality. The application scenarios of different thermal insulation systems and the possible problems of thermal design, system safety and fire prevention in these systems are given. This paper then puts forward the technical suggestions and the policy suggestions for the development of thermal insulation systems.

Keywords Carbon peaking and carbon neutrality goals; Thermal insulation systems for external wall; Application status; Engineering quality problem; Trend; Suggestion

基于光热性能模拟的中学教学楼形体设计研究

A Study on the Shape Design of Middle School Teaching Buildings Based on Simulation of Daylight and Thermal Performance

晏高亮，罗建河，赵立华，苏鑫雨，徐睿捷
Yan Gaoliang, Luo Jianhe, Zhao Lihua, Su Xinyu, Xu Ruijie

晏高亮，华南理工大学建筑学院，博士研究生。
罗建河，华南理工大学建筑设计研究院有限公司，高级工程师。
赵立华，华南理工大学建筑学院，教授。
苏鑫雨，华南理工大学建筑学院，硕士研究生。
徐睿捷，华南理工大学建筑学院，硕士研究生。

摘要 适应环境的形体设计对教学楼的光热环境有着关键性作用。目前校园建设用地紧张，教学楼形体复杂且功能复合化，这使得室内光热环境过度依赖主动式设备，造成高能耗现象。因此，有必要进行基于光热性能优化的建筑形体设计研究，以改善室内光热环境并降低能耗。本文以室内光热舒适度模拟为切入点，对参数化设计平台生成的三种典型教学楼形体进行评测，并由此总结出针对体块组合、平面及立面要素的具体设计策略，该策略可提升教学楼室内环境质量，实现节能减排，并为教学楼优化设计提供科学指引。

关键词 建筑形体设计；光热性能模拟；中学教学楼

Abstract The shape design that adapts to the environment plays a key role in the daylight and thermal environment of teaching buildings. Due to the shortage of land for campus construction, teaching buildings are complex in form and complex in function, which makes the indoor daylight and thermal environment over-rely active equipment, resulting in high energy consumption. Therefore, it is necessary to study on the building shape design based on corresponding performance optimization to improve the indoor daylight and thermal environment and reduce energy consumption. This study focuses on indoor daylight and thermal comfort and energy consumption simulation, evaluates three typical teaching building shapes generated by the parametric design platform, and summarizes the specific design strategies of block combination, plane form and elevation elements, which can improve the indoor environmental quality of teaching buildings, realize energy saving and emission reduction, and provide scientific guidance for the optimal design of teaching buildings.

Keywords Building shape design; Daylight and thermal performance simulation; Middle school teaching buildings

---

硬脂酸 – 泡沫铜复合相变材料的制备及其热性能研究

Preparation and Thermal Properties Study of Stearic Acid-Copper Foam Composite Phase Change Material

陈廷森，刘淑丽 *
Chen Tingsen, Liu Shuli

陈廷森，北京理工大学，博士研究生。
刘淑丽，北京理工大学，教授，博士生导师。

摘要 在建筑墙体中使用含相变材料（PCM）的蓄热层储存太阳能可降低建筑能耗。然而，PCM 的低导热性限制了储热速率，不利于建筑节能。泡沫铜具有质量轻、比表面积大、导热性好等优点，因此，本文以泡沫铜为多孔介质，采用真空吸附法制备了硬脂酸 - 泡沫铜复合相变材料并通过实验研究其热性能。研究表明，硬脂酸 - 泡沫铜复合相变材料与纯硬脂酸相比，储热过程熔化速率提升了 61.42%，放热过程凝固速率提升了 20%。硬脂酸 - 泡沫铜复合相变材料有利于太阳能的快速存储与释放，实现低碳节能建筑。

关键词 相变材料；建筑能耗；太阳能；泡沫铜；传热

Abstract The use of phase change materials (PCM) in thermal energy storage layers in building walls can reduce building energy consumption by storing solar energy. However, the low thermal conductivity of PCM limits the heat storage rate and is not conducive to building energy conservation. Copper foam has the advantages of being lightweight, and having large specific surface area, and good thermal conductivity. Therefore, this study used copper foam as a porous medium and employed the vacuum adsorption method to prepare a stearic acid-copper foam composite phase change material (CPCM), and experimentally studied its thermal properties. Results show that compared with pure stearic acid, the CPCM has a melting rate increased by 61.42% in the heat storage process and a 20% solidification rate in the exothermic process. The CPCM is conducive to the rapid storage and release of solar energy, and realizes low-carbon energy-saving buildings.

Keywords PCM; Building energy; Solar energy; Copper foam; Heat transfer

长三角地区考古遗产保护性建筑热工参数要求研究

Study on Thermal Parameter Requirements of Archaeological Heritage' s Protected Buildings in the Yangtze River Delta Region

胡雪晴，谢华荣，贺希皓，李永辉 *

Hu Xueqing, Xie Huarong, He Xihao, Li Yonghui

胡雪晴，东南大学建筑学院，硕士研究生。
谢华荣，东南大学建筑学院，助理研究员。
贺希皓，南京大报恩寺遗址博物馆，助理馆员，保管部主任。
李永辉，东南大学建筑学院，副教授，博士生导师。

摘要 目前，考古遗产保护存在着保护不当和运行成本高的问题。本文旨在明确基于长三角地区考古遗产保护性建筑的热工参数与文物本体劣化风险、建筑冷热负荷之间的量化关联。本文以南京大报恩寺遗址博物馆天王殿为研究对象，首先利用 EnergyPlus 软件计算大报恩寺天王殿室内热湿环境，并通过现场监测数据对模型进行验证。然后设置不同的外墙及玻璃幕墙的传热系数，得到不同工况下的室内温差及建筑能耗。结果表明，为满足文物保护和节能要求，应尽可能选用传热系数小的墙体保温材料。

关键词 遗产保护；保存环境；环境控制；劣化；节能

Abstract Presently, archaeological heritage conservation suffers from inappropriate protection and high operating costs. This paper aims to clarify the quantitative correlation between the thermal parameters of archaeological heritage' s protected buildings in the Yangtze River Delta region, the risk of deterioration of cultural relics, and the cooling and heating load of buildings. Taking Tianwang Hall of Nanjing Grand Bao'en Temple Site Museum as the research object, the EnergyPlus software was used to calculate the indoor hygrothermal environment of Tianwang Hall, and verify the model through on-site monitoring data. Then this paper sets the heat transfer coefficients for different facades and glass curtain walls to obtain the indoor temperature difference and energy consumption under different working conditions. The results show that to meet the requirements of heritage conservation and energy saving, wall insulation materials with small heat transfer coefficients should be used as much as possible.

Keywords Heritage conservation; Preservation environment; Environmental control; Deterioration; Energy saving

---

基于全生命周期碳排放的外墙保温层厚度适宜性分析

Suitability Analysis for Thickness of Facade Thermal Insulation Based on Life Cycle Carbon Emission

杨寒羽，姜召彩，孙立新，杨玉忠

Yang Hanyu, Jiang Zhaocai, Sun Lixin, Yang Yuzhong

杨寒羽，中国建筑科学研究院有限公司，助理工程师。
姜召彩，中国建筑科学研究院有限公司，工程师。
孙立新，中国建筑科学研究院有限公司，研究员，中国建筑科学研究院碳中和研究院副总工程师。
杨玉忠，中国建筑科学研究院有限公司，建筑安全与环境国家重点实验室，研究员，主任。

摘要 外墙保温层厚度的增加可降低建筑运行阶段的采暖空调能耗及相关碳排放，但过厚的保温层也会增加建筑建材生产阶段的碳排放。因此，外墙保温层如何在保障节能的同时减少碳排放有待进一步研究。本文以全生命周期碳排放为控制目标，基于能碳双控思路，选择 6 种典型外墙保温材料，采用数值模拟方法，对哈尔滨、北京、上海、广州的居住建筑全生命周期碳排放量进行模拟计算，给出基于减碳目标的各气候区典型保温材料厚度的适宜区间，为各气候区建筑外墙设计提供支撑。

关键词 外墙保温；全生命周期；碳排放；能碳双控

Abstract Increasing the thermal insulation thickness of the façade can reduce the HVAC energy consumption and related carbon emissions in the building operation phase, but also increases the carbon emissions in the production phase of building materials due to the excessively thick insulation. Therefore, it is to be studied how the facade insulation can reduce carbon emissions while guaranteeing energy efficiency. In this paper, we adopt the principle of energy and carbon dual control to achieve the control of the life cycle carbon emission. Six typical facade insulation materials are selected as research objects, and numerical simulation is employed to calculate the life cycle carbon emissions of residential buildings in Harbin, Beijing, Shanghai and Guangzhou. As a result, the desirable thermal insulation thickness of typical insulation materials in each climate zone is determined based on carbon emission reduction, which provides support for the design of building facades in each climate zone.

Keywords Facade thermal insulation; Life cycle; Carbon emission; Energy and carbon dual control

基于 Ladybug 的居住建筑界面改造设计研究

Research on Retrofit Design of Residential Building Interface Based on Ladybug

余海波，张辉，高宁呈，
金诗雨，黄晓玺
Yu Haibo, Zhang Hui,
Gao Ningcheng, Jin Shiyu,
Huang Xiaoxi

余海波，湖北工业大学土木建筑与环境学院，硕士研究生。
张辉，湖北工业大学土木建筑与环境学院，副教授，硕士生导师，新加坡国立大学访问学者。
高宁呈，湖北工业大学土木建筑与环境学院，硕士研究生。
金诗雨，湖北工业大学土木建筑与环境学院，硕士研究生。
黄晓玺，湖北工业大学土木建筑与环境学院，硕士研究生。

摘要 改造老旧小区以打造和谐宜居环境已成为重要的内容，对满足人民群众对美好生活的需求具有积极意义。本文以武汉新安九如片区的居住建筑为研究对象，运用 Rhinoceros&Grasshopper 建模工具，并借助 Ladybug 计算工具对建筑的室外环境进行模拟研究，发现建筑在立面窗户安装利用太阳能的潜力可达 79.2%。结合模拟结果，本文提出了可行的居住建筑界面改造设计思路和方法，以供城市更新中老旧小区改造借鉴和参考。

关键词 老旧小区；Ladybug；物理环境；居住建筑

Abstract Renovation of old neighborhoods to create a harmonious and livable environment has become one of the important elements, which is of positive significance in meeting people' s demand for a better life. The paper takes the residential building in the Wuhan Xin'an Jiuru area as the research object, uses Rhinoceros & Grasshopper modeling tools, and simulates the outdoor environment of the building with the help of Ladybug calculation tools, and finds that the building has the potential of solar energy installations for the use of the windows on the façade of the building which can be used by up to 79.2%. Combined with the simulation results, this paper proposes feasible design ideas and methods for the transformation of residential building interfaces for reference in the transformation of old neighborhoods in urban renewal.

Keywords Old neighborhoods; Ladybug; Built environment; Residential building

---

湿热地区光伏板表面水膜蒸发冷却对其热电性能的影响

Effect of Evaporative Cooling of Water Film on the Thermo-electrical Performance of Photovoltaic Panels in Hot and Humid Areas

徐杭，冉茂宇
Xu Hang, Ran Maoyu

徐杭，华侨大学建筑学院、厦门市生态建筑营造重点实验室，博士研究生。
冉茂宇，华侨大学建筑学院、厦门市生态建筑营造重点实验室，博士，教授。

摘要 光伏板表面采用水膜冷却是实现其降温增电的较好方式。然而，现有研究尚未从全年和季度考察光伏板受水膜冷却的降温增电效果。本文首先建立了光伏板水膜蒸发冷却传热模型，然后以湿热地区厦门为例，数值模拟光伏板上、下单面和双面水膜蒸发对光伏板全年温度和发电的影响。论文研究结论对于揭示和评估湿热地区光伏板水膜冷却性能有重要意义。

关键词 光伏板；蒸发冷却；发电效率；数值模拟

Abstract Water film cooling improves PV panel performance by reducing temperature and increasing electricity generation. However, current research lacks a comprehensive analysis of the year-round and quarterly effects of water film cooling. This study develops a heat transfer model and conducts simulations in Xiamen to evaluate the impact of water film cooling on PV panel temperature and electricity generation. The findings are crucial for understanding and optimizing water film cooling in humid and hot regions.

Keywords Photovoltaic panel; Evaporative cooling; Power generation efficiency; Numerical simulation

垂直绿化系统中蒸发蒸腾效应的综述

A Review of Evapotranspiration Effects in Vertical Greening Systems

余蓝星，张磊，张玉
Yu Lanxing, Zhang Lei, Zhang Yu

摘要 垂直绿化系统（VGS）通过遮阳、蒸发蒸腾、隔热为建筑物提供降温冷却。在目前的研究中，蒸发蒸腾作用对 VGS 降温效果的贡献仍然不清楚。为了填补该研究的空白，本研究对 VGS 蒸腾过程的降温效果进行了综述，概述了 VGS 的蒸腾现象以及蒸腾的降温效益；此外，对 VGS 蒸腾速率的监测方法和评估模型进行了讨论；最后，基于对 VGS 蒸散发研究的综合分析，提出了三点建议：对比研究不同 VGS 系统的蒸散发效益，进一步研究土壤体积含水量对 VGS 蒸散发的影响，通过实测数据验证并简化蒸散发理论模型。

关键词 垂直绿化系统；蒸发蒸腾；冷却效果；潜热通量

Abstract Vertical Greening Systems (VGS) provide cooling for buildings through shading, evapotranspiration, and insulation. In the literature available, the contribution of evapotranspiration to the cooling effect of VGS is still unclear. To fill this research gap, this study reviews the cooling effect of the evapotranspiration process of VGS. The transpiration phenomenon of VGS and the cooling benefits of transpiration are summarized. In addition, the monitoring methods and assessment models for the transpiration rate of VGS are also discussed in this review. Finally, based on the comprehensive analysis of VGS evapotranspiration research, three suggestions are put forward: a comparative study of evapotranspiration benefits of different VGS systems, further research on the impact of soil volumetric water content on VGS evapotranspiration, and verification and simplification of evapotranspiration through measured data.

Keywords Vertical Greening Systems(VGS); Evapotranspiration; Cooling effect; Latent heat flux

余蓝星，华南理工大学，硕士研究生。
张磊，华南理工大学，副研究员。
张玉，华南理工大学，助理研究员。

含盐多孔材料的等温吸湿和放湿性能试验研究

Experimental Study on Isothermal Hygroscopic and Dehumidifying Properties of Porous Materials Containing Salt

张宇，任鹏，孟庆林
Zhang Yu, Ren Peng, Meng Qinglin

摘要 沿海建筑围护结构受到盐雾的侵袭，因此其多孔材料含有盐分。对于含盐多孔材料储湿特性，可用等温吸放湿曲线进行表征。本文以红砂岩为例，构建 4 种含盐量、8 种湿度工况、吸放湿 2 种过程，测得平衡含湿量，绘制等温吸放湿曲线并提出拟合公式。以对照组（不含盐）和实验组（含盐）进行对比分析，结果表明，盐分对多孔材料的平衡吸放湿含湿量、吸放湿曲线产生的时滞效应均有促进作用，且含盐量越大，效果越明显。本研究可为含盐工况下多孔材料热湿盐耦合模型及其性能优化提供基础数据支撑。

关键词 含盐多孔材料；热湿盐耦合；等温吸放湿曲线；预测模型

Abstract Coastal building envelopes are subjected to salt spray and therefore their porous materials contain salt. The moisture storage characteristics of salt-containing porous materials can be characterized by isothermal moisture absorption and release curves. In this paper, we take red sandstone as an example, construct 4 kinds of salt content, 8 kinds of humidity conditions, and 2 kinds of moisture absorption and release processes, measure the equilibrium moisture content, draw isothermal moisture absorption and release curves and propose the fitting equations. Comparative analysis of the control group (without salt) and the experimental group (with salt) shows that salt has a promoting effect on the equilibrium moisture content of the porous material and the time lag effect produced by the moisture absorption and release curve; and the larger the salt content is, the more obvious the effect is. This study can provide basic data support for the heat-moisture-salt coupling model of porous materials and its performance optimization under salty working conditions.

Keywords Porous materials containing salt; Coupling of heat, moisture and salt; Isothermal moisture absorption and release curve; Prediction model

张宇，重庆大学建筑城规学院，博士研究生。
任鹏，华南理工大学建筑学院，教授。
孟庆林，华南理工大学建筑学院，教授。

建筑垂直绿化系统的降温节能机制

Mechanisms Behind Cooling and Energy-Saving Effects of Vertical Greenery Systems in Buildings

苏梅芳，石邢
Su Meifang, Shi Xing

摘要 建筑垂直绿化系统（VGS）将植被覆盖在建筑外立面上，通过冷却建筑物来减轻建筑能耗，其降温节能机制主要表现在遮阴、蒸散、保温和风障作用。本文首先梳理了 VGS 热效应的研究方法，其次总结了 VGS 在各项机制的表现及影响因素，研究发现夏季 VGS 的遮阴系数一般为 0.0 ～ 0.2，同时使墙体有效热阻增加 0 ～ 0.7 $m^2 \cdot K/W$。各机制效果受到 VGS 类型、植被和基质特征以及室外气象条件等因素的影响。最后，研究比较了各机制对整体热效应的贡献度，发现遮阴的贡献度约为蒸散的 2 倍。

关键词 垂直绿化系统；建筑；机制；方法；降温节能效应

Abstract Vertical greenery systems (VGS) integrate vegetation into the building façade to cool the building, thus reducing building energy consumption. Mechanisms behind cooling and energy-saving effects of VGS are mainly performed in shading, evapotranspiration, thermal insulation and wind barrier. This study first sorted out the methods for investigating the thermal effects of VGS. Secondly, the performance and influencing factors of VGS in each mechanism were summarized. It was found that the shading coefficient of VGS during summer ranged from 0.0 to 0.2, and they increased the effective thermal resistance of walls by 0~0.7 $m^2 \cdot K/W$. The effects of each mechanism are influenced by factors such as the type of VGS, vegetation and substrate characteristics, and outdoor weather conditions. Finally, the contributions of each mechanism to the overall thermal effects were compared, revealing that shading had a contribution approximately twice that of evapotranspiration.

Keywords Vertical greenery systems (VGS); Buildings; Mechanisms; Methods; Cooling and energy-saving effect

苏梅芳，同济大学，博士研究生。
石邢，同济大学建筑与城市规划学院，高密度人居环境生态与节能教育部重点实验室，教授，副院长 / 主任。

折线形特朗勃墙供暖性能影响因素的数值研究

Numerical Study on the Influencing Factors of Heating Performance of a Wavy-shape Trombe Wall

陈鸿宽，刘淑丽
Chen Hongkuan, Liu Shuli

摘要 特朗勃墙是一种被广泛应用在建筑中的被动式太阳能结构。折线形特朗勃墙在保留了传统特朗勃墙结构简单优势的同时，能够发挥比传统特朗勃墙更好的功能。本研究选取了 4 个影响折线形特朗勃墙供暖性能的核心参数进行研究，包括墙体折角、太阳高度角、太阳方位角和太阳辐照度。采用正交试验设计法选取了 16 个实验组，并使用计算流体力学方法进行数值仿真。结果表明，4 个因素对供热功率的影响从强至弱的排名为太阳方位角、太阳辐照度、太阳高度角、墙体折角，且对于最重要的设计参数墙体折角，其最佳取值为 95°。

关键词 建筑供暖；特朗勃墙；太阳能热利用；正交试验设计

Abstract The Trombe wall is a passive solar system that is widely used in buildings. The wavy-shape Trombe wall can achieve a better heating performance than the traditional Trombe wall, while retaining the advantage of the simple structure. In this study, 4 key factors influencing the heating performance of the wavy-shape Trombe wall were selected: the intersection angle of the wall, solar altitude angle, solar azimuth angle and intensity of the solar radiation. Then, 16 sets of experiments were designed based on the orthogonal design of experiments, and numerical simulations were conducted using computational fluid dynamics methods. The results showed that the influence of 4 factors on the heating power is ranked from the strongest to the weakest as follows: solar azimuth angle, intensity of the solar radiation, solar altitude angle, and intersection angle. Moreover, the optimal value of the design parameter intersection angle is 95°.

Keywords Building heating; Trombe wall; Solar thermal application; Orthogonal design of experiments

陈鸿宽，北京理工大学，博士研究生。
刘淑丽，北京理工大学机械与车辆学院，教授，博士生导师。

复合冰光热物性测试及对冰壳建筑热环境影响的初步探索

Test of Optical and Thermal Properties of Composite Ice and Preliminary Exploration of Their Effects on the Thermal Environment of Ice Shell Buildings

张媛媛，刘京，夏赟，武岳，田野，王震
Zhang Yuanyuan, Liu Jing, Xia Yun, Wu Yue, Tian Ye, Wang Zhen

张媛媛，哈尔滨工业大学，博士研究生。
刘京，哈尔滨工业大学，教授。
夏赟，黑龙江省寒地建筑科学研究院，研究员，副院长。
武岳，哈尔滨工业大学，教授。
田野，黑龙江省寒地建筑科学研究院，主任。
王震，哈尔滨工业大学，硕士研究生。

摘要 冰壳建筑是严寒及寒冷地区冰雪旅游产业中的重要项目。作为建筑施工建造的关键材料，复合冰的光热物理性质对于建筑围护结构传热性能及内部热环境特征影响显著。然而针对复合冰材料光热物理性质的研究尚属空白。本研究制备了不同比例纸浆纤维掺杂的复合冰材料，测试得到了光谱透射率、光谱反射率、导热系数、比热容、蓄热系数等光热物性参数，利用 EnergyPlus 软件模拟分析了不同材料组成冰壳建筑围护结构表面的热量传递及建筑内部空气温度，供实际工程设计和运行参考。

关键词 复合冰；光物性；热物性；围护结构

Abstract Ice shell building is an important project in ice and snow tourism in severely cold and cold regions. As a key material for its construction, the optical and thermal physical properties of composite ice significantly affect the heat transfer performance of the building envelope and the characteristics of the internal thermal environment. However, research on the optical and thermal physical properties of composite ice materials is lacking. In this study, composite ice materials doped with different proportions of pulp fibers were prepared, and optical and thermal physical parameters e.g. spectral transmittance, spectral reflectance, thermal conductivity, specific heat capacity, and heat storage coefficient were obtained through testing. The heat transfer on the surface of the building envelope and indoor air temperature of the ice shell buildings composed of different envelope materials were simulated and analyzed using EnergyPlus, providing a reference for the design and operation of the ice shell building in actual engineering.

Keywords Composite ice; Optical properties; Thermal properties; Building envelope

---

现代夯土建筑热性能研究进展

Advances in the Research of Thermal Performance of Modern Rammed Earth Buildings

王正阳，宋德萱，刘益
Wang Zhengyang, Song Dexuan, Liu Yi

王正阳，同济大学建筑与城市规划学院，博士研究生。
宋德萱，同济大学建筑与城市规划学院，教授，博士生导师。
刘益，同济大学建筑与城市规划学院，博士研究生。

摘要 夯土建筑是一种颇具环境潜力的建筑形式。相较于传统经验中“冬暖夏凉”的感性认知，定量研究更能揭示其性能原理，并为设计提供更加可靠的科学依据。本文阐述了夯土建筑热性能定量研究的材料学基础，从传热机理角度明确指出热湿耦合机制对现代夯土建筑环境调节的重要性，在此基础上对材料改性、构造改良的相关研究进行了归纳整理，最后总结了当前研究中的常见问题，并展望了未来的研究方向。

关键词 夯土建筑；热性能；热湿耦合；材料改性；构造改良

Abstract Rammed earth construction is an environmentally promising form of building. Quantitative research can reveal its performance principles and provide a more reliable scientific basis for design compared to the sensory perception of “warm in winter and cool in summer” in traditional experience. From the perspective of heat transfer mechanisms, it explicitly highlights the significance of the heat and moisture coupling mechanism in regulating the environmental conditions of modern rammed earth buildings. Moreover, it provides an overview of research related to material modification and structural improvement. Finally, the paper concludes by summarizing common issues in current research and offering prospects for future research directions.

Keywords Rammed earth buildings; Thermal performance; Heat and moisture coupling; Material modification; Construction improvement

多层相变材料组合对光伏板冷却及发电性能的初步探究

Preliminary Study on Cooling and Power Generation Performance of Photovoltaic Panels by Multilayer Phase Change Material Combination

韦登辉，冉茂宇
Wei Denghui, Ran Maoyu

摘要 光伏板工作温度升高严重影响了光伏板发电效率及其寿命，将相变材料用于光伏板降温因材料相变时吸放大量潜热，可较好地抑制其温度升高。然而，单一相变材料只在室外气候变化的某些时段有效。本文将不同相变温度的相变材料置于铝容器中，以不同层次组合贴附于光伏背板，测试了不同相变层组合对光伏板温度及发电效率的影响。结果表明：PCM 光伏板可降低光伏板峰值温度 10.59 ℃，平均降温 3.56 ℃；提高了输出功率 8.1%，发电效率平均提高了 0.46%。

关键词 光伏板；多层相变材料；降温；发电效率

Abstract The increase in the working temperature of photovoltaic panels seriously affects the power generation efficiency and life of photovoltaic panels, and the phase change material is used for cooling photovoltaic panels, because a large amount of latent heat is absorbed and released during the phase change of the material, which can better inhibit its temperature rise. However, single phase change materials can only be effective during certain periods of outdoor climate change. In this paper, phase change materials with different phase change temperatures are placed in aluminum containers and attached to photovoltaic back sheets at different levels, and the effects of different phase change layer combinations on the temperature and power generation efficiency of photovoltaic panels are tested. The results show that the PCM photovoltaic panel can reduce the peak temperature of the photovoltaic panel by 10.59 ℃, and the average temperature drop is 3.56 ℃. The output power is increased by 8.1%, and the power generation efficiency is increased by 0.46% on average.

Keywords Photovoltaic panel; Multilayer phase change material; Cooling effect; Power generation efficiency

韦登辉，华侨大学建筑学院，硕士研究生。
冉茂宇，华侨大学建筑学院，教授。

工作时间设定对办公建筑动态采光评价指标的影响

Impact of Working Time Settings on Dynamic Daylighting Evaluation Indicators in Office Buildings

何意，欧阳金龙，高庆龙
He Yi, Ouyang Jinlong, Gao Qinglong

摘要 动态采光评价指标是分析建筑采光性能的重要指标，与工作时间设定有密切关联。我国横跨五个时区，若以动态采光评价指标评价非东八区办公建筑的采光设计效果，应综合考量当地太阳时和天然光条件设定其工作时间。以东六区拉萨市四间办公室为例，依据当地采光条件设定合理的工作时区间，采用 Daysim 模拟分析动态采光评价指标。结果显示：动态模拟和评价办公建筑采光时，须结合当地采光条件设定天然采光时间，符合当地太阳时的工作时间设定有利于更高效地利用天然光，节省照明能耗。

关键词 时区；当地太阳时；天然采光；动态采光评价指标

Abstract Dynamic daylighting evaluation indicators are important indicators for analysing the daylighting performance of buildings and are closely related to the working time setting. Given that China spans 5 time zones, if the dynamic daylighting evaluation indicators are used to evaluate the lighting design effect of office buildings in non-GMT+8 Time Zone, the working hours should be set by taking into account the local solar time and natural light conditions. Taking four offices in Lhasa City, GMT+6 Time Zone as an example, a reasonable working hour interval is set according to the local lighting conditions, and the Daysim simulation is used to analyse the dynamic daylighting evaluation indicators. The results show that, when dynamically simulating and evaluating the lighting of office buildings, the natural lighting time must be set in combination with the local lighting conditions, and the working time set in accordance with the local solar time is conducive to the more efficient use of natural light and the saving of lighting energy consumption.

Keywords Time zone; Local solar time; Natural daylighting; Dynamic daylighting evaluation indicators

何意，四川大学，硕士研究生。
欧阳金龙，四川大学，教授。
高庆龙，四川大学，教授。

基于经济成本分析的珊瑚砂外墙保温层厚度确定

Determination of Coral Sand Insulation Thickness based on Economic Cost Analysis

高圣迪，谢静超 *，徐鑫，刘加平
Gao Shengdi, Xie Jingchao, Xu Xin, Liu Jiaping

摘要 极端热湿气候区具有高温高湿强辐射等气候特点，需要重点考虑墙体隔热，选用合理经济的外围护结构将关系到建筑的整体空调能耗。在不破坏当地生态环境的前提下，将珊瑚砂和海水制作成可用于建筑外墙中的珊瑚砂砌块。本文区分空调连续和间歇运行两种工况，综合考虑建筑空调运行能耗和保温材料在使用寿命内的经济成本，来优化双目标下的珊瑚砂砌块外墙的保温层厚度。结果表明：空调连续运行的最佳保温层厚度为 30 mm，投资回收期为 6.95 年；间歇运行的最佳保温层厚度为 20 mm，投资回收期为 7.77 年。

关键词 建筑外墙；投资回收期；保温层最佳厚度；珊瑚砂砌块

Abstract Extreme hot-humid climate zone has the climate characteristics of high temperature, high humidity and strong radiation, so it is necessary to pay attention to the wall insulation. The selection of a reasonable and economical outer protective structure relates to the overall air-conditioning energy consumption of the building. Coral sand and seawater are made into coral sand blocks that are used in building facades without damaging the local ecology. In this paper, two conditions of continuous and intermittent operation of air conditioning are distinguished, and the operating energy consumption of building air conditioning and the economic cost of thermal insulation materials within the service life are considered comprehensively to optimize the insulation layer thickness of the exterior wall of coral sand block under the dual objective. The results show that the optimum thickness of the insulation layer is 30mm and the payback period is 6.95 years. The optimum insulation thickness for intermittent operation is 20mm, and the payback period is 7.77 years.

Keywords Exterior wall; Investment payback period; Optimum insulation thickness; Coral sand block

高圣迪，北京工业大学，硕士研究生。
谢静超，北京工业大学，教授，北京工业大学绿色建筑环境与节能技术北京市重点实验室副主任。
徐鑫，北京工业大学，博士研究生。
刘加平，北京工业大学绿色建筑环境与节能技术北京市重点实验室，院士。

数据中心围护结构热工需求及分区研究

The Study on Thermal Requirements and Zoning of Data Center Envelope Structure

王礼钢，白雪莲
Wang Ligang, Bai Xuelian

摘要 与普通的民用建筑相比，数据中心应当考虑 IT 设备的全年散热问题，如果以“保温隔热”的思想来设计围护结构，反而不能降低其空调系统用能需求。本研究结合我国数据中心主要布局城市，在现行标准规范的基础上，模拟计算了当围护结构热工参数变化时在不同气候条件下数据中心的全年空调负荷变化情况，分析了热工参数敏感性，探讨了数据中心的围护结构热工性能需求，从而提出了相应的围护结构热工设计分区及数据中心围护结构设计要求建议。研究结果可为数据中心围护结构的设计优化提供参考。

关键词 数据中心；围护结构；热工性能；能耗模拟；分区设计

Abstract Compared with ordinary civil buildings, by data centers one should consider the year-round heat dissipation of IT equipment, and if the envelope structure is designed with the idea of “thermal insulation”, it cannot reduce its air conditioning energy demand. This study focuses on the main cities with data centers in China, based on the current standards and codes, simulates and calculates the year-round air conditioning load changes of data centers under different climatic conditions when the thermal parameters of the envelope structure change, analyzes the sensitivity of the thermal parameters and discusses the thermal performance requirements of the envelope structure of the data centers, to put forward the recommendations for corresponding enclosure thermal design zoning and data center enclosure design requirements. The results of the study can provide a reference for the design optimization of data center envelope structures.

Keywords Data center; Envelope structure; Thermal performance; Energy consumption simulation; Zoning design

王礼钢，重庆大学土木工程学院，硕士研究生。
白雪莲，重庆大学土木工程学院，教授，博士生导师。

装配式轻钢龙骨复合外墙热工性能实验研究

Experimental Study on Thermal Performance of Prefabricated Light-gauge Steel Framing Composite External Walls

王金霞，易嘉，胡宴桥
Wang Jinxia, Yi Jia, Hu Yanqiao

摘要 轻钢龙骨复合外墙作为一种新型的装配式外墙，具有轻质、装配化程度高等优点，提升其热工性能对“双碳”目标下的建筑节能具有重要意义。本文采用防护热箱法对 8 种不同构造组成的装配式轻钢龙骨复合外墙进行热工实验研究，分别从传热系数、传热阻的理论值与实测值方面进行对比分析。结果表明：轻钢龙骨空腔内填充的保温材料、构件连接的绝热处理及附加的外保温装饰板，对提升墙体热工性能均具有较大作用，为装配式轻钢龙骨复合外墙的推广和应用提供了理论依据和设计参考。

关键词 轻钢龙骨复合外墙；热工性能；防护热箱法；传热系数

Abstract As a new type of prefabricated exterior wall, light-gauge steel framing composite exterior wall has the advantages of lightweight, high degree of assembly and so on. Improving the thermal performance of wall is of great significance for building energy-saving under the carbon peaking and carbon neutrality goals. In this paper, the thermal performance of prefabricated light-gauge steel framing composite external walls with eight different structures is measured by guard hot-box. The heat transfer coefficient, total thermal resistance, theoretical value and measured value are compared and analyzed. The results show that the thermal insulation material filled in the light-gauge steel framing composite external walls, the insulation treatment of the component connection and the additional external insulation decorative panel have a great effect on improving the thermal performance of walls. It provides a theoretical basis and design reference for the promotion and application of prefabricated light-gauge steel framing composite external walls.

Keywords Light-gauge steel framing composite external wall; Thermal performance; Guard hot-box; Heat transfer coefficient

王金霞，中建科技集团西部有限公司，初级工程师。
易嘉，中建科技集团西部有限公司，高级工程师。
胡宴桥，中建科技集团西部有限公司，初级工程师。

建筑能耗模拟中敏感性分析方法综述

A Review of Sensitivity Analysis Methods in Building Energy Modeling

李建锋，张磊，张玉
Li Jianfeng, Zhang Lei, Zhang Yu

摘要 敏感性分析在建筑能耗模拟中起着重要的作用，它可以从建筑能耗模拟中识别影响建筑能耗的关键变量。本文首先介绍了在建筑能耗模拟中如何确定输入变量以及模型的建立运行方法。其次，回顾了建筑能耗模拟中使用的敏感性分析方法，包括全局方法与局部方法，同时描述了不同敏感性分析方法在评估建筑能耗方面的优缺点。再次，总结了前人工作中所识别出的影响建筑能耗模拟的关键变量。最后，提出了未来进一步研究的建议，以便在评估建筑能耗时提供更稳健的分析。

关键词 建筑能耗模拟；敏感性分析；关键变量；全局方法；局部方法

Abstract Sensitivity analysis plays an important role in the building energy modeling, which can identify key variables that affect building energy consumption from the building energy modeling. The article first introduces how to determine input variables in building energy consumption simulation and the establishment and operation methods of the model. Secondly, the sensitivity analysis methods used in building energy analysis are reviewed, including global methods and local methods, and the advantages and disadvantages of different sensitivity analysis methods in evaluating building energy consumption are described. Thirdly, the key variables identified in previous work affecting building energy modeling are summarized. Finally, recommendations for further research in the future are presented to provide a more robust analysis when assessing building energy consumption.

Keywords Building energy modeling; Sensitivity analysis; Key variables; Global method; Local method

李建锋，华南理工大学建筑学院，硕士研究生。
张磊，华南理工大学建筑学院，副研究员。
张玉，华南理工大学化学与化工学院，助理研究员。

严寒地区农房太阳能热水地板采暖系统经济分析

Economic Analysis of Solar Hot Water Floor Heating System of Rural Houses in Severe Cold Areas

杜晨，张九红
Du Chen, Zhang Jiuhong

摘要 本文针对严寒地区农房太阳能热水地板采暖系统耗电量及系统设备造价等方面的使用成本及其经济可行性进行研究，以沈阳某农房为例，在围护结构热工性能及室内热环境温度满足当地设计标准前提下，为分析太阳能收益，选用能耗模拟软件 Design-Builder 对纯电能采暖方案下的采暖耗电量进行模拟，并与太阳能 + 电能综合采暖方案的实测数据进行耗电量对比，以此方法来预估该设备动态投资回收期。结果表明，该系统从使用成本经济性角度具有一定的推广前景，且太阳能在该采暖系统下对采暖能耗及经济性的影响可为同地区同类型建筑节能设计提供参考依据。

关键词 太阳能热水地板采暖系统；农房；太阳能收益；经济性

Abstract This paper aims to study the cost and economic feasibility of solar hot water floor heating system in rural houses in severe cold areas, taking a rural house in Shenyang as an example. Under the premise that the thermal performance of the enclosure structure and the temperature of the indoor thermal environment satisfy the local design standard, in order to analyze the solar energy gain, the energy simulation software Design-Builder is selected to simulate the power consumption of the heating under the pure electric energy heating scheme and compare the power consumption with that of the solar energy + electric energy integrated heating scheme. The power consumption is compared with the measured data of the combined solar + electric heating program, and the dynamic payback period of the equipment is estimated in this way.

Keywords Solar hot water floor heating system; Farm houses; Solar benefits; Economy

杜晨，东北大学，博士研究生。
张九红，东北大学，教授，博士生导师。

武汉高层住宅太阳能－空气源热泵供热系统的优化选择

Optimal Selection of Solar-Air Source Heat Pump Systems for High-Rise Residential Buildings in Wuhan

吕晨茜，王振 *
Lü Chenxi, Wang Zhen

摘要 基于推进新能源在建筑领域应用的发展需求，将分布式能源系统应用到住区是一种有利于节能减排的解决方案。本研究针对高层住宅，采用数值模拟与参数优化相结合的研究方法，首先通过 DeST 软件获取建筑负荷特性，其次结合 TRNSYS 软件对系统运行状况进行模拟，最后以费用年值最低、系统能效比最高为目标函数，采用 NSGA- Ⅱ算法进行多目标优化，并引入 TOPSIS 法获得系统最优决策解，提出实现负荷最优匹配的方案，为武汉高层住宅应用太阳能 - 空气源热泵提供参考。

关键词 高层住宅；太阳能－空气源热泵；能耗模拟；NSGA-II 算法；多目标优化

Abstract Based on the development need to promote the application of new energy in the building sector, the application of distributed energy systems to settlements is a solution conducive to energy conservation and emission reduction. The study is aimed at high-rise dwellings and uses a combination of numerical simulation and parameter optimization. The building load characteristics were first obtained through DeST. The system operating conditions were then simulated in conjunction with TRNSYS. Finally, the NSGA-II algorithm is used for multi-objective optimization with the lowest annual value of cost and the highest energy efficiency ratio of the system as the objective function, and the TOPSIS method is introduced to obtain the optimal decision solution of the system. The scheme to realize the optimal matching of loads is proposed to provide a reference for the application of solar-air source heat pumps in high-rise residential buildings in Wuhan.

Keywords High-rise residential building; Solar-air source heat pump; Energy consumption simulation; NSGA-II algorithm; Multi-objective optimization

吕晨茜，华中科技大学中欧清洁与可再生能源学院，硕士研究生。
王振，华中科技大学建筑与城市规划学院，副教授。

天空开阔度对建筑辐射制冷屋面的影响研究

Research on the Effect of Sky View Factor on Radiative Cooling Roofs on Buildings

陈维杰，李春莹，唐海达 *
Chen Weijie, Li Chunying, Tang Haida

陈维杰，深圳大学建筑与城市规划学院，深圳市医养建筑重点实验室（筹建启动），硕士研究生。
李春莹，深圳大学建筑与城市规划学院，深圳市医养建筑重点实验室（筹建启动），副教授，硕士生导师。
唐海达，深圳大学建筑与城市规划学院，深圳市医养建筑重点实验室（筹建启动），研究员，硕士生导师。

摘要 辐射制冷技术作为一种新型被动式节能技术，早期单独应用在建筑屋面，但会受到周边建筑遮挡从而降低辐射冷屋面的天空开阔度，进一步影响冷却性能。本研究在深圳地区实地测试，实验结果定量揭示了辐射制冷平均温降与天空开阔度的关系，随着辐射体的天空开阔度降低，其冷却性能降低的趋势并非呈线性关系。当天空开阔度从 0.7 下降到 0.1 时，辐射体的冷却性能下降了 60%。本研究为建筑辐射冷屋面在复杂城市环境中的冷却性能潜力提供参考。

关键词 被动式节能；辐射制冷；天空开阔度；辐射冷却性能；城市建筑

Abstract Radiative sky cooling (RSC) as a new passive energy-saving technology was initially applied to building roofs alone, but can be shaded by surrounding structures thereby reducing the sky view factor (SVF) for radiative cooling roofs, further affecting the radiant cooling performance. This research was field-tested in the Shenzhen area, the results show that the relationship between the mean temperature drop of a radiation cooler and SVF is quantified experimentally. As the SVF of the radiator decreases, radiant cooling performance does not tend to decrease in a linear manner. As SVF decreases from 0.7 to 0.1, the cooling performance of the radiation cooler decreases by 60%. This research provides a reference for the radiant cooling performance potential of radiative cooling roofs on buildings in complex urban environments.

Keywords Passive energy saving; Radiative sky cooling; Sky view factor; Radiant cooling performance; Urban architecture

---

课程思政在建筑物理课程中的教学探索与研究

Teaching Exploration and Research of Curriculum Ideology and Politics in the Course of Building Physics

王万江，刘香，樊辉
Wang Wanjiang, Liu Xiang, Fan Hui

王万江，新疆大学建筑工程学院，教授。
刘香，新疆大学建筑工程学院，硕士研究生。
樊辉，新疆大学建筑工程学院，讲师。

摘要 高校的核心任务是立德树人，在“授业”中“传道”，让知识承载价值，培养适合我国社会主义建设合格的接班人。课程思政兼顾智育与德育，如何将思政元素有机地融入专业知识传授之中，是课堂思政建设的探索点。本文以建筑物理课程为例，探讨了课程与课程思政深度融合的必要性。通过提取各章节中的思政元素，有机融入教学的全过程，以实现将如“盐”般的课程思政元素潜移默化、无声无息地融入“水”一样的专业知识，进而提高教学质量和学生职业素养，为今后推进高校专业课程思政教育提供参考和借鉴。

关键词 建筑物理；课程思政；教学方法；教学改革

Abstract The core mission of the university is to set up the moral education and cultivate young talents. To “teach” in “teaching”, to make knowledge carry value, and to train qualified successors for Chinese socialist construction. It is a new standard in curriculum ideology and politics to integrate curriculum ideology and politics into the teaching of professional knowledge. In this paper, the author discusses the necessity of integrating the curriculum with the curriculum ideology and politics, and draws out the political factors in every chapter to integrate them into the teaching process, so as to make the “salt” into the specialized knowledge. This research aims to improve the teaching quality and the professional level of the students, which will be used as a reference for the further development of higher vocational courses.

Keywords Building physics; Curriculum ideology and politics; Teaching methods; Teaching reform

夏热冬暖地区新型热致变色高反射涂料的制备研究

Preparation and Research of New Thermochromic High-Reflection Coatings in Hot Summer and Warm Winter Areas

刘萍，孟庆林，毛会军，汪博然
Liu Ping, Meng Qinglin, Mao Huijun, Wang Boran

刘萍，华南理工大学建筑学院，硕士研究生。
孟庆林，华南理工大学建筑学院，教授，亚热带建筑科学国家重点实验室副主任，中国建筑学会建筑物理分会理事长。
毛会军，华南理工大学建筑学院，博士研究生。
汪博然，华南理工大学建筑学院，硕士研究生。

摘要 本文结合热致变色涂料与高反射涂料的特性，制备了46种不同颜色与变色温度的热致变色高反射涂料，并分析了颜色、变色温度及掺混比例对混合涂料光谱反射比的影响。结果表明，与热致变色涂料相比，混合涂料变色前后的反射比差异更大，其中黑色与蓝色的混合涂料变色前后的反射比差异较大，可达23%。变色温度基本不影响变色前后的反射比差值。高反射涂料掺混比例越大，混合涂料变色前后的反射比差异越大。

关键词 热致变色涂料；高反射涂料；光谱反射比；变色温度；掺混比例

Abstract In order to optimize the performance of the existing thermochromic coatings, 46 kinds of thermochromic high-reflection coatings with different colors and color-change temperatures were prepared by combining the characteristics of thermochromic coatings and high-reflection coatings. The effects of color, color-change temperature and mixing ratio on the spectral reflectance of the mixed coatings were analyzed. The results show that compared with thermochromic coatings, mixed coatings exhibit greater differences in reflectance before and after degradation, with the reflectance difference of black and blue mixed coatings being the greatest (up to 23%). The color-change temperature does not affect the reflectance difference. The greater the proportion of high-reflection coating, the greater the reflectance difference before and after color change.

Keywords Thermochromic coatings; High-reflection coating; Spectral reflectance; Color-change temperature; Mixing ratio

---

复合相变墙体简化计算方法适用范围的研究

Study on the Applicable Range of Simplified Calculation Method for Composite Phase Change Wall

王乙晴，白璐 *
Wang Yiqing, Bai Lu

王乙晴，内蒙古工业大学土木工程学院，硕士研究生。
白璐，内蒙古工业大学土木工程学院，讲师。

摘要 将蓄热相变材料应用于复合墙体中，能够起到稳定室温和降低采暖热负荷等作用，明确复合相变墙体简化计算方法的适用范围，对于相变材料在建筑中的应用有重要意义。对复合相变墙体建立两种传热计算模型，即简化模型与传统模型，对其蓄热过程进行数值模拟，分析模型差异性，得到简化模型的适用范围。结果表明：随着普通层厚度增加，两种方法差异性增大；单位面积比热可作为衡量简化方法适用性的指标；当最大差异系数小于15%，单位面积比热小于63.4 kJ/(m^2·K)时，墙体温度分布可以用简化方法来计算。

关键词 相变材料；复合墙体；简化模型；蓄热过程

Abstract Applying thermal storage phase change materials to composite walls can stabilize room temperature and reduce heating load, clarifying the applicability of simplified calculation methods for composite phase change walls, which is of great significance for the application of phase change materials in buildings. Two heat transfer calculation models were established for composite phase change walls, namely the simplified model and the traditional model, and their heat storage process was numerically simulated, the differences between the models were analyzed, and the applicable range of the simplified model was obtained. The results indicate that as the thickness of the ordinary layer increases, the difference between the two methods increases; The specific heat per unit area is proposed as an index to measure the applicability of the simplified method. When the maximum difference coefficient is less than 15% and the specific heat per unit area is less than 63.4 kJ/(m^2 · K), the wall temperature distribution can be calculated by the simplified method.

Keywords Phase change materials; Composite wall; The simplified model; Heat storage process

中国住宅生活热水系统应用现状研究

Research on the Application Status of Residential Domestic Hot Water System in China

田帅，周欣，石邢，金星
Tian Shuai, Zhou Xin, Shi Xing, Jin Xing

摘要 生活热水系统的形式对住宅能耗有着重要的影响。为了解中国住宅生活热水系统的应用现状，本研究收集了来自全国的 726 份有效问卷，调研了江苏省 57 个绿色住宅项目，并对其生活热水系统进行深入分析。结果显示，目前国内住宅建筑的生活热水系统以分散式电热水器和燃气热水器为主，占比分别为 45.5% 和 40%，太阳能热水系统占比为 12.4%。分散式太阳能热水系统是绿色住宅建筑中应用可再生能源的最主要的技术手段，其中，太阳能生活热水系统以屋顶安装集热器的形式为主，占比 91.3%。在补热方面，大多数绿色住宅项目选择电作为补热能源。

关键词 生活热水系统；太阳能热水系统；绿色住宅；问卷调研

Abstract The form of domestic hot water system has an important influence on residential energy consumption. To investigate the application status of domestic hot water system in China, 726 valid questionnaires were collected. At the same time, this study surveyed 57 green residential districts in Jiangsu Province, and thoroughly analyzed the application of their domestic hot water systems. The results show that the domestic hot water system in residential buildings is dominated by decentralized electric water heaters and gas water heaters, accounting for 45.5% and 40% respectively, while solar water heating systems account for 12.4%. The distributed household solar water heating system is the most important measure to apply renewable energy in green residences. The solar collector is mainly mounted on the roof, accounting for 91.3% of the total solar water heating systems. Most of the green residential districts choose electricity as the replenishing energy source of solar water heating system.

Keywords Domestic hot water system; Solar water heating system; Green residence; Questionnaire survey

田帅，同济大学建筑与城市规划学院，博士研究生。
周欣，东南大学建筑学院，副教授。
石邢，同济大学建筑与城市规划学院，教授。
金星，东南大学建筑学院，教授。

建筑物理仿真模拟实验在建筑物理教学中的实践与探讨

Practice and Discussion of Building Physics Simulation Experiment in Building Physics Teaching

王万江，杨威
Wang Wanjiang，
Yang Wei

摘要 建筑物理实验在课程教学中有实验多、种类杂、周期长的特点，存在对设备操作熟练度要求较高、多数精密仪器依赖教师、操作学生重视程度低等问题，对建筑物理仿真模拟实验的开发迫在眉睫。本研究通过搭建建筑物理仿真模拟实验教学平台，从了解设备使用方式到熟悉操作，进行数据处理并完成实验报告等步骤来优化教学。通过重构教学内容，贯彻“模拟实验 + 实践操作”同频共振的教学方法。试用表明，该方式可提高物理实验教学效果，一定程度上解决当前建筑物理实验教学困境，对促进培养综合性人才具有重要的现实意义。

关键词 建筑物理；模拟仿真；教学模式

Abstract Building physics experiment in the teaching phase have the characteristics of many experiments, miscellaneous stypes and long cycles, and there are problems such as high requirements for equipment operation proficiency, reliance on teachers for most precision instruments, and low attention of operating students,etc., so the development of simulated analogue experiments in building physics is extremely urgent. This study optimizes the teaching by constructing a teaching platform for building physics simulation experiment, from understanding how to use the equipment to being familiar with the operation, processing data and completing the experimental report and other steps. By reconstructing the teaching content and implementing the teaching method of “simulation experiment + practical operation”, the trial shows that this method can improve the effect of physics experiment teaching, solve the current dilemma of building physics experiment teaching to a certain extent, and have important practical significance in promoting the cultivation of comprehensive talents.

Keywords Building physics; Simulation; Teaching mode

王万江，新疆大学建筑工程学院，教授。
杨威，新疆大学建筑工程学院，硕士研究生。

基于 CFD 模拟的外呼吸式双层幕墙节能优化研究

Energy Saving Optimization Study of Externally Ventilation Double Skin Facade on CFD Simulation

钟凯洁，刘琦
Zhong Kaijie, Liu Qi

摘要 双层自然通风幕墙在热压作用下产生浮升力来实现空腔内部通风的效果，并提升其在实际工程使用中的热舒适性和节能性。本研究通过 CFD 模拟分析热通道内的温度分布、流动特性等要素，探讨了外呼吸式双层自然通风幕墙在不同季节工况、不同朝向等情况下的热工性能，提出了适应当地气候条件且具有良好热工性能的双层幕墙构造，以及双层自然通风幕墙在环境因素、使用条件差异下的节能优化策略。

关键词 双层通风幕墙；自然通风；烟囱效应；节能幕墙

Abstract The use of NVDSF can make use of thermal pressure to achieve the effect of internal ventilation of the cavity and improve its thermal comfort and energy saving in actual engineering use. This study analyses the thermal performance of externally ventilated NVDSF under different seasonal conditions and different orientations through CFD software simulations. By analysing the temperature distribution and flow characteristics in the hot channel, the study proposes a NVDSF structure with good thermal performance that is adapted to the local climatic conditions, and an energy saving strategy for NVDSF under different environmental factors and usage conditions.

Keywords Double skin facade; Natural ventilation; Solar chimney effect; Energy saving facade

钟凯洁，山东建筑大学建筑城规学院，硕士研究生。
刘琦，山东建筑大学建筑城规学院，副教授。

温湿度对全珊瑚骨料－气凝胶混凝土导热系数的影响研究

A Study on the Effect of Temperature and Humidity on the Thermal Conductivity of Whole Coral Aggregate-Aerogel Concrete

刘康，王莹莹
Liu Kang, Wang Yingying

摘要 本研究采用将聚丙烯纤维和不同体积掺量的气凝胶颗粒掺入珊瑚砂混凝土的方法，制备出兼具良好力学性能与热性能的海水拌制全珊瑚骨料－气凝胶混凝土（CAAC），并对不同温湿度条件下材料的热湿物性参数进行试验研究。结果表明，CAAC 导热系数随温度变化的增长率为 4.0%~8.4%，随相对湿度变化的增长率为 21.5%~31.9%，随气凝胶体积掺量变化（0~45%）的下降率为 62.3%~82.0%。

关键词 珊瑚骨料混凝土；气凝胶颗粒；导热系数；不同温湿度条件

Abstract In this paper, by adding polypropylene fibers and aerogel particles of different volumes to coral sand concrete, the whole coral aggregate-aerogel concrete(CAAC) with good mechanical and thermal properties was prepared. Then, the thermal and wet physical parameters of the materials were investigated under different temperature and humidity conditions. The results show that the thermal conductivity of CAAC increases by 4.0%~8.4% with temperature, increases by 21.5%~31.9% with relative humidity, and decreases by 62.3%~82.0% with volume content of aerogel (0~45%).

Keywords Coral aggregate concrete; Aerogel particles; Thermal conductivity; Various hygrothermal environment

刘康，西安建筑科技大学，博士研究生。
王莹莹，西安建筑科技大学，教授。

建筑节能设计用气象数据的获取方法——以“一带一路”为例

Meteorological Data Acquisition Method for Building Energy Conservation Design: A Case Study of the Belt and Road

霍英博，李红莲，宋学锟，和玺，吕文
Huo Yingbo, Li Honglian, Song Xuekun, He Xi, Lü Wen

霍英博，西安建筑科技大学，硕士研究生。
李红莲，西安建筑科技大学，教授。
宋学锟，和玺，吕文，西安建筑科技大学，硕士研究生。

摘要 准确提供当地气象参数是评价建筑能耗的前提条件和基础工作，室外气象参数如何准确获取是亟待解决的问题。本文总结了气象数据获取几个关键问题的研究，并给出了相关方法，对“一带一路”地区气象资料现状进行了综述，做了以下工作：①回顾了气象参数初始数据的获取。②提出了无气象台站观测地区气象数据的获取方法。③对气温模型进行误差分析，验证气象数据预测的准确性。

关键词 气象参数；空间插值；数据精度

Abstract Accurate provision of local meteorological parameters is a prerequisite and basic work for evaluating the energy consumption of buildings. How to obtain outdoor meteorological parameters accurately is an urgent problem. This paper summarizes the research on several key problems of meteorological data acquisition and gives the relevant methods. The current situation of meteorological data in the Belt and Road region is summarized, and the following work is done: ① The initial data acquisition of meteorological parameters is reviewed. ② The method of obtaining meteorological data in the observation area without meteorological stations is proposed. ③ The error analysis of the temperature model was carried out to verify the accuracy of the meteorological data prediction.

Keywords Meteorological parameter; Spatial interpolation; Data accuracy

---

基于有限元的水泥基建筑材料有效导热系数预估方法研究

A Study on Prediction Method of Effective Thermal Conductivity of Cement-based Building Materials Based on Finite Element Method

陈威，王莹莹
Chen Wei, Wang Yingying

陈威，西安建筑科技大学，博士研究生。
王莹莹，西安建筑科技大学，教授。

摘要 水泥基建筑材料是建造现代建筑的最基本建材，其导热系数的准确确定对于建筑负荷及能耗计算至关重要。本文利用随机生长法重构了水泥基建筑材料的内部孔隙结构，通过有限元方法对水泥基建筑材料进行传热过程模拟及有效导热系数计算。本文进而研究了孔隙结构参数与水泥基材料导热系数之间的影响关系，结果显示，在孔隙率为 0.3 时，不同孔隙数量、不同孔径与不同孔隙规则程度造成的水泥基建筑材料导热系数的波动最剧烈。

关键词 水泥基材料；有效导热系数预估；随机生长；有限元

Abstract Cement-based building materials are the most basic building materials for the construction of modern buildings, and the accurate acquisition of their thermal conductivity is crucial for building load and energy consumption calculations. In this paper, the random growth method is used to reconstruct the internal pore structure of cement-based building materials, and the finite element method is used to simulate the process of heat transfer and calculate the effective thermal conductivity of cement-based building materials. The relationship between the pore structure parameters and the thermal conductivity of cement-based materials is studied. The results show that when the porosity is 0.3, the thermal conductivity of cement-based building materials caused by different pore number and pore size and different pore regularity degree fluctuates most acutely.

Keywords Cement-based materials; Prediction of effective thermal conductivity; Random growth; Finite element

可解释性机器学习揭示城市空间形态对建筑能耗的影响

Explainable Machine Learning Reveals the Impact of Urban Spatial Form on Building Energy Consumption

张鼎，张法卓，王锦煜，黄辰宇，姚佳伟
Zhang Ding, Zhang Fazhuo, Wang Jinyu, Huang Chenyu, Yao Jiawei

张鼎，张法卓，同济大学建筑与城市规划学院，本科生。
王锦煜，同济大学建筑与城市规划学院，硕士研究生。
黄辰宇，同济大学建筑与城市规划学院，博士研究生。
姚佳伟，同济大学建筑与城市规划学院，副教授，硕士生导师。

摘要　城市空间形态对城市建筑能源建模（UBEM）具有显著影响。本研究通过耦合热岛效应的 UBEM 引擎计算了北京市 1847 个住宅类型的城市街区建筑能耗，使用地理加权回归（GWR）分析街区建筑能耗的空间异质性，根据 GWR 回归系数对街区能耗进行空间聚类，使用可解释性机器学习 LightGBM 对各子类建模并使用 SHAP 输出特征重要性排序。结果显示建筑面积标准差与正面面积指数对街区建筑能耗影响最大，建筑间距与天空可视因子对建筑能耗影响在北京中心城区、西城区和东南城区的依赖关系不同。

关键词　城市空间形态；城市建筑能源建模；地理加权回归；可解释性机器学习

Abstract　The urban spatial form significantly influences Urban Building Energy Modeling (UBEM). In this study, the UBEM engine, coupled with the urban heat island effect, was used to calculate the building energy consumption of 1,847 residential-type urban blocks in Beijing. The spatial heterogeneity of block energy consumption was analyzed using Geographically Weighted Regression (GWR). Based on the GWR regression coefficients, block energy consumption was spatially clustered. LightGBM was employed to model each subclass, and SHAP was used to output the feature importance ranking. The results revealed that the standard deviation of building area and frontal area index had the greatest impact on block energy consumption. The relationship between building spacing and building height and their impact on energy consumption differed in the central, western, and southeast areas of Beijing.

Keywords　Urban spatial form; Urban building energy modeling; Geographically weighted regression; Explainable machine learning

---

豫中地区新农宅冬季建筑碳排放研究

Study on Winter Building Carbon Emissions of New Rural Residential Buildings in Central Henan Povince

周赛赢，夏博，唐山
Zhou Saiying, Xia Bo, Tang Shan

周赛赢，长安大学，硕士研究生。
夏博，长安大学，副教授。
唐山，长安大学，硕士研究生。

摘要　为了研究豫中农村地区新农宅建筑碳排放情况，采用问卷调查的方法收集各农宅信息，选取豫中地区漯河市舞阳县周马庄村典型新农宅作为研究对象，采用分项能耗记录的方法，收集该农宅冬季最冷月 1 月在取暖、生活用电、炊事等方面建筑用能数据，采用碳排放因子法计算各项数据的碳排放量。本研究发现该农宅在冬季取暖方面的碳排放最多，并进一步提出相应的改进策略。

关键词　新农宅；碳排放；能耗；取暖方式

Abstract　To study the carbon emissions of new rural residential buildings in rural areas of central Henan, questionnaires were used to collect information on rural residential buildings. The typical new rural residential buildings in Zhoumazhuang Village, Wuyang County, Luohe City, central Henan Province were selected as the research objects. The sub-item energy consumption recording method was used to collect the building energy data of the rural residential buildings in heating, domestic electricity, cooking and other aspects in the coldest winter month of January, and the carbon emission factor method was used to calculate the carbon emissions of various data. The study discovered that the farmhouse has the highest carbon emissions in winter heating, and then proposed corresponding improvement strategies.

Keywords　New rural residential buildings; Carbon emissions; Energy consumption; Heating method

基于风热耦合的下沉式商业建筑入口形态研究

Study on Sunken Entrance Space Form in Commercial Buildings Based on Wind and Thermal Environment Simulation

郭奕岑，陈言，赵庆卓，张海滨
Guo Yicen, Chen Yan, Zhao Qingzhuo, Zhang Haibin

摘要 为抵御夏季的高温热浪，商业建筑往往室内外温差很大。这时，从室内到室外的温度过渡显得就很重要。因此，需要对商业建筑入口空间形态进行设计研究，营造更为舒适的室内外过渡环境。本研究以重庆的下沉式商业建筑入口为例，通过现场调研，总结概括出该地区入口空间典型模型，并对典型模型进行风热环境模拟，观察入口空间形态变化对风热环境的影响情况，基于入口过渡区最佳流体场、温度场的目标，提出入口空间形态设计策略，为改善西南地区下沉式商业建筑入口的风热环境舒适度提供参考。

关键词 高温地区；商业建筑；下沉入口空间；空间形态；风热环境舒适度

Abstract With the goal of resisting high temperature, commercial buildings often create a large temperature gap between indoor and outdoor environments. Therefore, it' s crucial to shorten the temperature transition gap from indoor to outdoor by researching the design of the entrance vestibule, which can create a more comfortable transition environment. Based on the examples of commercial building sunken entrances in Chongqing, this study summarizes the typical entrance vestibule model in Chongqing through field surveys. It observes differences in wind and thermal environments when the entrance space form is altered. Combining the target of optimizing the fluid and thermal field in this transition area, the related design strategy of entrance space form is proposed, which provides a reference for improving the wind and thermal comfort of the sunken entrance space of commercial buildings in the southwest region of China.

Keywords Hot area; Commercial buildings; Sunken entrance vestibule space; Spatial form; Comfort of wind and thermal environment

郭奕岑，陈言，赵庆卓，重庆大学建筑城规学院，本科生。
张海滨，重庆大学建筑城规学院，副教授，建筑技术科学系副系主任。

厦门地区复合相变隔热屋面节能性研究

Study on the Energy Efficiency of the Composite Phase Change Insulation Roof in the Xiamen Area

胡惠玲，冉茂宇 *
Hu Huiling, Ran Maoyu

摘要 相变材料（PCM）应用于建筑屋面，是有效的屋面隔热手段。PCM在白天吸收太阳辐射能，降低屋面内表面温度同时衰减温度波。本文通过OpenStudio 联合 EnergyPlus 软件建立建筑模型，模拟分析在厦门气候环境下建筑屋面外表面温度在制冷期内分布情况，并选取相应 PCM，探究在屋面填充单一相变温度 PCM 及复合多种相变温度 PCM 对建筑制冷能耗的影响。研究结果表明：①屋面添加单一 PCM 能有效隔热，降低建筑制冷能耗，但仅在一定温度范围内发挥潜热。②与单一 PCM 相比，采用复合 PCM 能够有效降低建筑制冷能耗，建筑节能率提高 9%。

关键词 复合相变；隔热屋面；建筑能耗

Abstract The application of phase change materials (PCM) to building roofs is an effective means of roof insulation. PCM absorbs solar radiation energy during the day and reduces the inner surface temperature of the roof while attenuating temperature waves. In this paper, we build a building model through OpenStudio joint EnergyPlus software, simulate and analyze the distribution of the external surface temperature of the building roof during the cooling period under the climate environment of Xiamen, and select the corresponding PCM to investigate the effect of filling the roof with a single phase change temperature PCM and compounding multiple phase change temperature PCM on the building cooling energy consumption. The results show that ① the addition of single PCM can effectively insulate the roof and reduce the building cooling energy consumption, but it only exerts latent heat in a certain temperature range. ② Compared with single PCM, the use of composite PCM can effectively reduce the building cooling energy consumption and increase the building energy saving rate by 9%.

Keywords Composite phase change; Insulated roofing; Building energy

胡惠玲，华侨大学建筑学院，博士研究生。
冉茂宇，华侨大学建筑学院，教授。

冷屋顶的地域适用性与性能优化研究

Study on Regional Adaptability and Performance Optimization of Cool Roof

独孤俊，叶海，林贤哲，许其弘
TokGo Jun, Ye Hai, Rim Hyon Chol, Xu Qihong

摘要 冷屋顶是一种可以降低夏季制冷能耗的同时还可以缓解城市热岛效应的新型建筑结构形式。本文使用 Grasshopper 中 Ladybug 调用 Energyplus 的 CondFD 计算引擎，对小型低层办公建筑进行能耗模拟，评估了冷屋顶在中国各气候区的适用性，并对 PCM 冷屋顶进行了性能优化。研究结果显示冷屋顶只在夏热冬暖地区有明显的节能效果，在其他地区节能效果极小甚至反而导致全年能耗的增加，在夏热冬暖地区太阳辐射反射率从 0.1 变化至 0.9 时全年空调负荷可以减少 3.7%。优化结果，显示采用相变温度 22℃的 PCM 安装在保温层内侧时候效果最佳，此时 PCM 冷屋顶可以降低全年空调负荷 4.4%。

关键词 冷屋顶；相变材料；建筑节能；PCM

Abstract Cool roof is a new type of building structure that can reduce cooling energy consumption in summer and alleviate urban heat island effect at the same time. This paper uses Ladybug in Grasshopper to invoke the CondFD computing engine of Energyplus to simulate the energy consumption of small low-floor office buildings, evaluates the adaptability of cool roofs in different climate zones in China, and optimizes the performance of PCM cool roofs. The results show that the cool roof has an obvious energy-saving effect only in hot summer and warm winter areas, but in other areas the energy-saving effect is very small or even leads to an increase in annual energy consumption. In hot summer and warm winter areas, when the solar radiation reflectance changes from 0.1 to 0.9, the annual air-conditioning load can be reduced by 3.7%. The optimization results also show that the effect is best when PCM with a phase change temperature of 22°C is installed on the inside of the insulation layer, and the PCM cold roof can reduce the annual air-conditioning load by 4.4%.

Keywords Cool roof; Phase change material; Architectural energy saving; PCM

独孤俊，同济大学建筑与城市规划学院，硕士研究生。
叶海，同济大学建筑与城市规划学院，硕士生导师。
林贤哲，上海理工大学环境学院，硕士研究生。
许其弘，同济大学建筑与城市规划学院，硕士研究生。

我国严寒和寒冷地区热管房太阳能采暖潜力研究

Solar Heating Potential of Heat Pipe House in Severely Cold and Cold Regions of China

寇方铖，许同芯，王馨
Kou Fangcheng, Xu Tongxin, Wang Xin

摘要 新型热管太阳房利用平板重力热管的超高当量导热系数及热二极管性可大幅改善冬季室内热环境，实现零碳采暖。热管房采暖效果与其内围护结构热性能密切相关。本文对热管房在我国严寒和寒冷地区 61 个不同地点进行内围护结构参数优化，模拟计算其最佳被动采暖效果，分析了当地辐射温差比与热管房改善室内热不舒适的太阳能应用潜力的关系。结果表明：①热管房在大多数地点的最优内墙墙体材料导热系数和体积比热容均为常规建材取值范围内的最大值。②在所选地点改善室内热不舒适的程度在 29.9%~100%。③当地辐射温差比与热不舒适改善度呈高度线性正相关。

关键词 自然资源；太阳房；被动房；热二极管；围护结构

Abstract The novel heat pipe solar house can greatly improve the indoor thermal environment in winter and realize zero-carbon heating by exploiting the ultrahigh equivalent thermal conductivity and thermal diode property of the flat gravity heat pipe. In this paper, the indoor envelope parameter optimization and the passive heating effect of the heat pipe house are simulated at 61 locations in severely cold and cold regions of China. The results show that: ① The optimal thermal conductivity and volume specific heat capacity of the interior wall material of heat pipe house at most locations are the maximum values within the range of conventional building materials. ② The degree of improvement in indoor thermal discomfort at selected locations ranges from 29.9% to 100%. ③ There is a highly linear positive correlation between the ratio of radiation to temperature difference.

Keywords Natural resources; Solar house; Passive house; Thermal diode; Building envelope

寇方铖，清华大学建筑学院，博士研究生。
许同芯，清华大学建筑学院，本科生。
王馨，清华大学建筑学院，副研究员。

石灰基秸秆保温墙体热工性能研究

Research on Thermal Performance of Lime-based Straw Concrete Envelope

杨丽平，刘衍，郑佳瑞，崔临风，杨柳
Yang Liping, Liu Yan, Zheng Jiarui, Cui Linfeng, Yang Liu

杨丽平，西安建筑科技大学建筑学院，博士研究生。
刘衍，西安建筑科技大学建筑学院，教授。
郑佳瑞，西安建筑科技大学建筑学院，硕士研究生。
崔临风，西安建筑科技大学建筑学院，硕士研究生。
杨柳，西安建筑科技大学建筑学院，教授。

摘要 我国西藏地区传统民居外围护结构普遍采用单一建筑材料砌筑，冬季室内空气温度较低，石灰基秸秆混凝土是性能优良的绿色环保材料，具有导热系数低、低碳等特点，可提高围护结构保温性能。本文采用谐波反应法计算对比西藏地区现有民居建筑外墙与石灰基秸秆砌块墙体的衰减延迟、净失热热流和内表面温度。结果表明，石灰基秸秆砌块墙体失热热流小，墙体内表面温度较现有墙体高 3.2~5.3℃。

关键词 围护结构；石灰基秸秆砌块；谐波反应法；热流

Abstract In Xizang, the indoor air temperature of the regional building is low due to the single material of the envelope. Lime-based straw concrete is a kind of green material with low thermal conductivity, low carbon and other characteristics, which can improve the thermal insulation performance of the envelope. The harmonic response method was used to calculate and compare the attenuation delay, net heat loss flux and internal surface temperature of the residential building wall and the lime-based straw concrete wall in Xizang. The results show that the heat loss of the lime-based straw block wall is small, and the internal surface temperature of the wall is 3.2~5.3℃ higher than the existing wall.

Keywords Envelope; Lime-based straw block; Harmonic response method; Heat flux

---

花岗岩表面温度和粗糙度对发射率影响的实验研究

Investigating the Effect of Granite Surface Temperature and Roughness on Emissivity

夏静雯，任鹏
Xia Jingwen, Ren Peng

夏静雯，华南理工大学建筑学院，亚热带建筑科学国家重点实验室，硕士研究生。
任鹏，华南理工大学建筑学院，亚热带建筑科学国家重点实验室，副教授，硕士生导师。

摘要 发射率作为遥感获取地表温度和研究地表能量平衡的重要参量，易受到材料表面物性的影响，其准确性会严重影响热环境分析的结果。为解决此问题，基于傅里叶变换热红外和光栅光谱法，测量了 3 种花岗岩表面在不同温度和粗糙度时的发射率。研究结果表明，3 种花岗岩表面在 8~14 μm 波段的发射率均随温度升高而减小，下降幅度在 0.019~0.028，在 9.8~14 μm 波段更为显著。而其表面在 8~14 μm 波段的发射率随粗糙度增加而增大，变化增量超过 0.02，8.5~10 μm 波段为变化特征波段。

关键词 表面发射率；温度；粗糙度；热红外光谱；花岗岩

Abstract Land surface emissivity constitutes a crucial parameter in the remote sensing inversion of land surface temperature and investigations of surface energy balance, that is easily affected by the surface state of the material, and its accuracy will seriously affect the thermal environment analysis results. In order to solve this problem, based on Fourier transform infrared and grating spectroscopy, the surface emissivities of three granites at different temperatures and roughness were measured. The results show that the surface emissivity of the three granites in the 8~14 μm band decreases with the increase of temperature, and the decrease range is between 0.019~0.028, and it is more significant in the 9.8~14 μm band. The emissivity of its surface in the 8~14 μm band increases with the increase of roughness, and the change increment exceeds 0.02, and the 8.5~10 μm interval is the characteristic band of change.

Keywords Surface emissivity; Temperature; Roughness; Thermal infrared spectra; Granite

垂直绿化空间布局对行人感知与景观偏好的影响

Effect of Spatial Arrangement of Vertical Greening Systems on Pedestrians' Environmental Perception and Landscape Preference

沈小洁，杨峰 *
Shen Xiaojie, Yang Feng

沈小洁，同济大学建筑与城市规划学院，博士研究生。
杨峰，同济大学建筑与城市规划学院，博士，副教授，博士生导师。

摘要 紧凑集约的城市化进程带来了极具压迫感的高密度街道景观，给宜人宜居城市环境创建和居民身心健康发展带来挑战。垂直绿化系统（VGS）作为补偿高密度城区绿地损失、提升城市景观品质的有效手段，在改善居民情绪、压力等方面的潜力有待深入研究。借助虚拟现实工具开展模拟实验，自我报告与可穿戴生理监测设备获得的心理、生理数据被用于探讨垂直绿化空间布局对行人景观偏好、情绪状况和压力缓冲的影响。研究结果将为量化评估 VGS 的健康效益以及更全面、可持续的 VGS 规划设计提供证据。

关键词 垂直绿化系统；景观偏好；环境感知；情绪；缓冲效应

Abstract Intensive urbanization has resulted in oppressive high-density streetscapes, which are an obstacle to achieving a livable and pleasant urban environment and a threat to the health of urban dwellers. The Vertical Greening System (VGS) has been supported as an effective approach to compensate for the loss of green space in dense urban areas and to improve the quality of the urban landscape, with the potential for stress reduction and emotion improvement. Through simulation experiments in visual reality, data generated from self-report measures and wearable biosensing measures are used to explore the relationship between the spatial arrangement of VGS and pedestrians' landscape preference, perceived emotion and stress. The findings provide evidence for a quantitative assessment of the health benefits of VGS and for a more comprehensive and sustainable VGS design.

Keywords Vertical greening system; Landscape preference; Environment perception; Emotion; Buffering effect

预制混凝土夹心保温墙体连接件局部热桥效应研究

Study on Local Thermal Bridge Effect of Precast Concrete Sandwich Insulation Wall Connectors

郑佳瑞，刘衍，杨丽平，崔临风，杨柳
Zheng Jiarui, Liu Yan, Yang Liping, Cui Linfeng, Yang Liu

郑佳瑞，西安建筑科技大学建筑学院，硕士研究生。
刘衍，西安建筑科技大学建筑学院，教授。
杨丽平，西安建筑科技大学建筑学院，博士研究生。
崔临风，西安建筑科技大学建筑学院，硕士研究生。
杨柳，西安建筑科技大学建筑学院，教授。

摘要 预制混凝土夹心保温墙体由内外叶混凝土墙板、中间保温层及连接件组成。由于连接件的导热系数与墙体主体部分相差较大，容易产生热桥效应，因此，本文选取导热系数较大的不锈钢连接件作为研究对象，采用理论分析与数值模拟的方法对不锈钢桁架式连接件的局部热桥效应进行研究。研究发现：不锈钢连接件的热桥效应随着墙体保温性能的提升而更加显著，当使用挤塑聚苯板为保温材料时，随着保温层厚度从 180 mm 增加到 340 mm，连接件的存在使墙体平均传热系数增大了 5.01%~5.51%。

关键词 预制夹心保温墙体；连接件；热桥；热工性能

Abstract The precast concrete sandwich insulation wall is composed of inner and outer leaf concrete wall panels, intermediate insulation layer and connecting parts. Because the thermal conductivity of the connector is much different from that of the main part of the wall, it is easy to produce thermal bridge effect. Therefore, this paper selects stainless steel connectors with large thermal conductivity as the research object, adopts theoretical analysis and numerical simulation to study the degree of local thermal bridge influence of stainless steel truss connectors, and it found that the thermal bridge effect of stainless steel connector increased with the improvement of wall insulation performance, and when extruded polystyrene board was used as the insulation material, the thickness of the insulation layer increased from 180mm to 340mm, and the existence of the connector increased the heat transfer coefficient of the wall by 5.01%~5.51%.

Keywords Precast sandwich insulation wall; Connectors; Thermal bridge; Thermal performance

湿热地区透水铺装蒸发降温特性研究

Study on Evaporative Cooling Characteristic of Pervious Pavements in Hot and Humid Areas

汪俊松，孟庆林，谭康豪，祁乾龙
Wang Junsong, Meng Qinglin, Tan Kanghao, Qi Qianlong

摘要 透水铺装可利用水分蒸发降温带走热量，改善城市热环境。然而，科学评估手段的缺乏导致实际工程中其降温性能差异较大。本文明确了单面浸泡实验用于评价透水铺装材料吸水及保水性能的有效性；研究了造成其蒸发降温差异的原因，发现吸水速率对其高速蒸发过程有着重要贡献；通过夏季室外实验探明了其对室外热环境的影响，发现透水铺装最大可降低表面温度10℃，可使近地面热环境指标 WBGT 最大降低 2℃。

关键词 室外热环境；透水铺装；吸水及保水特性；蒸发降温特性

Abstract Pervious pavements could not only infiltrate water to reduce the surface runoff, but also cut off a certain amount of water and utilize water evaporation to take the heat away, thus improving the urban thermal environment. However, due to the lack of assessment methods, the difference of evaporative cooling performance of different types of pervious pavements was pretty large in practice. In this study, the effectiveness of partial immersion test as an evaluation of water absorption and retention performance of pervious pavements were clarified. Afterwards, reasons that lead to the difference of evaporative cooling performance have been explored, and the water absorption rate was found to be a critical factor affecting the high-speed evaporation process. Moreover, the outdoor experiments in summer further reveal that pervious pavement can reduce the surface temperature by 10℃ , and the outdoor thermal index WBGT by up to 2℃ , respectively.

Keywords Outdoor thermal environment; Pervious pavements; Hydrological characteristic; Evaporative cooling characteristic

汪俊松，华南理工大学建筑学院，博士后助理研究员。
孟庆林，华南理工大学建筑学院，教授。
谭康豪，华南理工大学材料科学与工程学院，博士后助理研究员。
祁乾龙，华南理工大学建筑学院，博士研究生。

以近零能耗为导向的高层住宅光伏利用研究与发展综述

Review on Research and Development of Photovoltaic Utilization for High-rise Residential Buildings Based on Nearly Zero-Energy Consumption

袁唐鹏赟，邓丰 *
Yuan Tangpengyun, Deng Feng

摘要 以近零能耗为导向，太阳能光伏的利用是高层住宅实现其运行阶段能源供给平衡的关键。本文综述了国内外近零能耗高层住宅在光伏利用方面的研究，探讨了全球近零能耗建筑的发展趋势以及光伏技术在住宅建筑中的应用现状，总结了光伏利用在高层住宅中的能源优势和潜在问题，并详细探讨了光伏组件的选型、安装方式、安装倾角和位置、发电效率和投资回报、能源储存与转化，以及光伏系统与住宅外立面的矛盾等问题。最后，本文总结了光伏在高层住宅中的应用前景和研究方向。

关键词 近零能耗；高层住宅；光伏利用；住宅立面

Abstract Guided by the concept of nearly Zero-energy consumption, the utilization of solar photovoltaics (PV) is crucial for achieving energy supply balance during the operational phase of high-rise residential buildings. This paper provides a comprehensive review of the research on PV utilization in domestic and international nearly Zero-energy consumption high-rise residential buildings. It explores the global trends in nearly Zero-energy buildings and the current status of PV technology in residential construction. The paper summarizes the energy advantages and potential issues of PV utilization in high-rise residential buildings, and delves into various aspects such as PV component selection, installation methods, installation tilt angle and position, electricity generation efficiency and return on investment, energy storage and conversion, as well as the conflict between PV systems and residential facades. Finally, it concludes by summarizing the prospects and research directions of PV application in high-rise residential buildings.

Keywords Nearly Zero-energy consumption; High-rise residential buildings; Photovoltaic utilization; Residential facade

袁唐鹏赟，同济大学建筑与城市规划学院，硕士研究生。
邓丰，同济大学建筑与城市规划学院，副教授。

基于空调系统的高效机房现状及其实现途径研究

Research on the Current Situation and the Ways to High Efficiency of Central Air-conditioner Room Based on Air Conditioning System

刘双胜
Liu Shuangsheng

摘要 空调是建筑用能的大户，空调机房又是空调用能最多的部分，因而其节能和减碳潜力巨大。我国空调机房在设计、模拟、施工、运营等各方面都存在诸多问题，高效机房的实现任重而道远。本文基于我国空调机房现状，从标准制定、软件模拟、设备升级、优化设计、控制策略等方面指出实现高效机房的有效途径和意义。

关键词 高效机房；能效；优化设计；控制策略

Abstract The energy consumption of air conditioning is the major energy consumption of buildings, and it is the most carbon emission part of building operation, so its energy saving and carbon reduction potential is huge. There are many problems in the design, simulation, construction, and operation in our country' s central air-conditioner rooms, and the realization of efficient rooms is still a long way to go. Based on the current situation of air-conditioner rooms in China, this paper points out the effective ways and significance of realizing efficient rooms from the aspects of standard formulation, software simulation, equipment upgrading, optimization design and control strategy.

Keywords High efficiency central air-conditioner room; Energy efficiency; Optimal design; Control strategy

刘双胜，四川大学建筑与环境学院，硕士研究生。

超高层建筑在烟囱效应作用下的压差特性模拟研究

Simulation Study on Differential Pressure Characteristic of Stack Effect on Super High-rise Buildings

冒慧
Mao Hui

摘要 超高层建筑在冬季寒冷天气条件下烟囱效应明显，由此会引起电梯门开闭故障、啸叫、渗风能耗增加等一系列问题。本文以某栋 400 m 高的超高层建筑为案例，使用建筑多区域网络模型计算软件 CONTAM 对该建筑的烟囱效应进行模拟，分析了超高层建筑在冬季烟囱效应作用下的压差分布状况，并讨论了建筑构件的渗透特性、内部隔断设置和首层出入口形式对电梯门两侧热压差的影响效果。结果显示，增设内部隔断对降低电梯门两侧热压差效果最明显，其次为增加建筑幕墙的气密性等级。

关键词 超高层建筑；烟囱效应；数值模拟；压差分布

Abstract The stack effect of super high-rise buildings is obvious in cold weather, which could cause a series of problems, such as the failure of elevator door opening and closing, howling noise, and increased energy consumption due to air seepage. Taking a 400m tall super high-rise building as an example, this paper simulates the stack effect of the super high-rise building using the building multi-zone network model computer program CONTAM, analyzes the pressure distribution of stack effect on super high-rise buildings, and discusses the influence of multiple factors such as the permeability characteristics of building components, the internal partition setting and the form of entrance on the thermal pressure difference between the two sides of elevator doors. The results show that the most effective measure to reduce the pressure difference of the elevator shaft is adding internal partitions, followed by increasing the air tightness level of the building curtain wall.

Keywords Super high-rise building; Stack effect; Numerical simulation; Pressure distribution

冒慧，上海市建筑科学研究院有限公司，技术工程师。

110 kV 户内变电站通风散热模拟及优化研究

Simulation and Optimization of Ventilation and Heat Dissipation in 110 kV Indoor Substation

刘诗雨
Liu Shiyu

摘要 本文针对户内变电站通风不畅、空调能耗较高的缺点，以主要设备功能房间为研究对象，分析并优化了机械通风系统相较于空调系统的散热及节能效果。运用 CFD 数值模拟技术，针对设备位置及房间特点，对进排风口位置影响下的气流组织进行了模拟分析；对多种通风工况下的散热效果进行了模拟分析，结合全年环境温度探讨了机械通风的节能潜力。

关键词 户内变电站；通风散热；气流组织；节能减碳

Abstract This paper is discussing the analysis and optimization of mechanical ventilation systems for main equipment rooms, in response to the issues of poor ventilation and high energy consumption of air conditioning systems in indoor substations. Using CFD technology, the airflow organization under the influence of inlet and outlet positions is simulated and analyzed based on equipment location and room characteristics. The heat dissipation effects under various ventilation conditions are also simulated and analyzed, and the energy-saving potential of mechanical ventilation is explored in conjunction with an analysis of the ambient temperature throughout the year.

Keywords Indoor substation; Ventilation and heat dissipation; Airflow organization; Energy saving and carbon reduction

刘诗雨，上海市建筑科学研究院有限公司，工程师。

蓄热通风建筑热工设计基础研究

Research on the Basic Problems of Thermal Design of Thermal Storage and Night Ventilation Buildings

乔宇豪，刘衍，杨柳 *
Qiao Yuhao, Liu Yan, Yang Liu

摘要 蓄热通风技术是在建筑本体良好蓄热性能基础上，利用夏季夜间室外凉空气对房间进行降温，是我国西北等大陆性气候主导下地区建筑实现零空调能耗的理想途径。本文针对蓄热通风技术气候适宜性分析方法、蓄热通风建筑外墙热反应特性等基础问题开展研究，提出“气候降温潜力度时比”指标，划定我国蓄热通风技术适宜性分区。基于正交实验方法，开展人工控制环境下的缩尺模型实验，揭示蓄热通风过程下的外墙热反应特性并提炼关键热工性能参数。

关键词 蓄热通风技术；气候适宜性分析；外墙热反应特性；缩尺模型实验；热工性能参数

Abstract Thermal storage and night ventilation technology is a passive cooling solution based on good thermal storage performance, utilizing the cool outdoor air of summer night to decrease the indoor air and interior surface temperature. It is an ideal way to achieve zero air-conditioning energy consumption in continental climate dominated areas. In this paper, the basic problems of thermal design are investigated, including climate suitability analysis method of thermal storage and night ventilation technology, and thermal characteristics of thermal storage and night ventilation building exterior walls. The indicator named “climate cooling potential degree-hour ratio” is proposed, and the suitability zone in China is defined. Based on the orthogonal experimental method, the reduced-scale experiments under the artificially controlled condition are carried out to reveal the thermal characteristics of exterior walls, and the key thermal performance parameters are refined.

Keywords Thermal storage and night ventilation technology; Climate suitability analysis; Thermal characteristics of exterior wall; Reduced-scale experiment; Thermal performance parameters

乔宇豪，西安建筑科技大学交叉创新研究院，助理教授。
刘衍，西安建筑科技大学建筑学院，教授。
杨柳，西安建筑科技大学建筑学院，教授。

内嵌毛细管适应性墙体采暖工况热性能实验研究

Experimental Study of Multilayer Wall with Thermal Capillary Network in Heating Condition

赵田，庄智*，郭海新，项平
Zhao Tian, Zhuang Zhi, Guo Haixin, Xiang Ping

摘要 内嵌毛细管适应性墙体是一种能够利用自然低品位能源并根据环境变化动态可调的新型围护结构。本文设计并搭建墙体测试平台，研究采暖工况下有无内热源及进水温度的变化对于墙体热性能的影响。研究发现：①毛细管热源层是一道良好的室内外热屏障，内嵌毛细管适应性墙体的热性能优于常规墙体；②水温的升高改善了墙体的热工性能和舒适性。不同进水温度下内壁面谷值温度相较常规墙体分别提升 0.27℃、1.31℃ 和 2.04℃，内壁面日平均散热损失热流密度分别降低 26.66%、95.17% 和 173.82%。

关键词 低品位能源；内嵌毛细管适应性墙体；节能潜力；热舒适性

Abstract The multilayer wall with thermal capillary network (MTC embedded wall) is a new type of envelope that can utilize natural low-grade energy and be dynamically adjusted according to the change of environment. In this paper, a MTC embedded wall test platform was designed and built to study the influence of inner heat source and water temperature on the thermal performance of the wall in heating conditions. The results show that: ① Thermal capillary network layer is a good indoor and outdoor thermal barrier, and the comprehensive performance of MTC embedded wall is better than that of conventional wall; ② The increase of water temperature improves the energy saving potential of the wall and indoor thermal comfort at the same time. Compared with the conventional wall, the valley temperature of the inner wall under different inlet water temperature increases by 0.27℃, 1.31℃ and 2.04℃, respectively. The daily mean heat flux decreases by 26.66%, 95.17% and 173.82%, respectively.

Keywords Low-grade energy; MTC embedded wall; Energy-saving potential; Thermal comfort

赵田，同济大学建筑与城市规划学院，博士研究生。
庄智，同济大学建筑与城市规划学院/同济大学中德工程学院，副教授。
郭海新，德国柏林工业大学德中科技文化交流中心，教授。
项平，同济大学土木工程学院，研究员。

保温结构一体化外墙热工设计问题分析及优化研究

The Research of the Thermal Design for the Integration Wall with Thermal Insulation

张昭瑞，赵亮，姜召彩，孙立新
Zhang Zhaorui, Zhao Liang, Jiang Zhaocai, Sun Lixin

摘要 “双碳”目标下保温结构一体化外墙体系成为外墙保温行业重要发展方向，目前标准体系尚未完善。本文将对现有保温结构一体化外墙体系及其热工问题进行梳理、归纳，以点联式内置保温现浇混凝土体系为研究对象，通过模拟计算的方式进行热工性能量化分析，提出热工设计优化相关建议。

关键词 保温结构一体化墙体；连接固定件；热工性能；热工设计

Abstract With the proposal of carbon peaking and carbon neutrality goals, the integration wall with thermal insulation has become an important development direction of external wall insulation, and the current relevant standard provisions are not perfect. In this paper, the existing integration wall will be sorted out and summarized, the thermal performance analysis of the integration wall connected by steel tube will be carried out through the way of simulation calculation, and the thermal design optimization suggestions will be proposed.

Keywords The integration wall with thermal insulation; Connector; Thermal performance; Thermal design

张昭瑞，建科环能科技有限公司，高级工程师。
赵亮，建科环能科技有限公司。
姜召彩，建科环能科技有限公司。
孙立新，中国建筑科学研究院有限公司，正高级工程师。

基于重力平板热管的光伏被动房热电性能研究

Thermo-electric Performance of a Photovoltaic Building Based on Gravity Flat Heat Pipe

刘羿铭，寇方铖，刘宇舟，王馨
Liu Yiming, Kou Fangcheng, Liu Yuzhou, Wang Xin

摘要 将 BIPV 建筑与具有高导热系数和热二极管性的平板重力热管结合，可以在降低光伏板温度、提高发电效率的同时，使光伏废热能够传递进入内墙，达到被动供热的目的。本文建立了一种基于平板热管的光伏被动房的数值模型并进行了实验验证；比较了玻璃罩对该光伏热管房的热电性能的影响，并分析了不同气候区内该光伏热管房的热电性能。研究表明，在光伏电池外侧增加玻璃罩可使北京供暖季典型日的平均室温相比无玻璃罩的光伏热管房提高 6.0℃；在所计算的 5 个北方典型气候城市中，光伏热管房在敦煌能够提供相对舒适的室内热环境，室内平均温度达到 12.5℃，比光伏对照房高 3.8℃，比无光伏对照房高 2.9℃。

关键词 太阳能建筑；BIPV/T；蓄放热；热二极管；当量导热系数

Abstract Combining BIPV buildings with flat heat pipes with high thermal conductivity and thermal diode properties can reduce the temperature of photovoltaic panels and improve power generation efficiency, while enabling photovoltaic waste heat to conduct into the inner wall, and achieve for passive heating purposes. In this paper a numerical model of a photovoltaic passive house based on flat heat pipes is established and verified by experiments, and the influence of the glass outer window on the thermos-electric performance of the photovoltaic heat pipe house is compared, while the thermos-electric performance of the photovoltaic heat pipe house in different climate zones is analyzed. It is discovered that adding glass covers to the outside of photovoltaic cells can increase the average room temperature on a typical day in Beijing' s heating season by 6.0℃ compared with photovoltaic heat pipe houses without glass covers. Among the 5 typical cities calculated, photovoltaic heat pipe houses in Dunhuang can provide relatively comfortable indoor thermal environment, with an average indoor temperature of 12.5 ℃ , which is 3.8℃ higher than the control room and 2.9℃ higher than the non-photovotaic room.

Keywords Solar building; BIPV/T; Heat storage and release; Thermal diode; Equivalent thermal conductivity

刘羿铭，清华大学，硕士研究生。
寇方铖，清华大学，博士研究生。
刘宇舟，清华大学，本科生。
王馨，清华大学，副研究员。

基于建筑能源应用的多能互补系统研究现状

Research Status of Multi-energy Complementary System Based on Building Energy Application

傅芷涵，季亮，陈晓菲，刘佳鹭
Fu Zhihan, Ji Liang, Chen Xiaofei, Liu Jialu

摘要 “双碳”背景下，本文基于找寻最优化的建筑能源系统这一目标，首先，说明了当前多能互补系统的研究现状，介绍了各项关键技术。其次，综述了电、燃气和热能这三种能源互相搭配形成的多种混合储能技术，介绍了每种储能系统的组成、特点及运行策略；针对上述多种储能设备及系统，基于建筑电力负荷特性，介绍了各能源系统容量配置运行优化和模拟研究方法。最后，介绍了多能互补能源系统实用性与经济性对比分析研究，为今后基于建筑能源应用的最优化多能互补系统的研究提供了参考。

关键词 建筑能源系统；混合储能；光储直柔；运行优化

Abstract Under the background of carbon peaking and carbon neutrality goals, based on the objective of finding the optimal building energy system, this paper firstly explains the current research status of multi-energy complementary systems and introduces the key technologies. Secondly, the paper reviews the hybrid energy storage technologies of electricity, gas and thermal energy, and introduces the composition, characteristics and operation strategies of each energy storage system; then, the capacity configuration and operation optimization of each energy system based on the building electric load characteristics are presented. Finally, the paper presents a comparative analytical study of the practicality and economics of multi-energy complementary energy systems, which provides reference for the future research of the optimal multi-energy complementary systems based on building energy applications.

Keywords Building energy system; Hybrid energy storage; Operation optimization

傅芷涵，上海市建筑科学研究院有限公司，初级工程师。
季亮，上海市建筑科学研究院有限公司，教授级高级工程师，绿色节能与低碳创新研究院副院长。
陈晓菲，上海市建筑科学研究院有限公司，初级工程师。
刘佳鹭，上海市建筑科学研究院有限公司，初级工程师。

夏热冬冷地区民居类建筑遗产的能耗现状分析——以一栋传统合院建筑为例

Analyzing Energy Performance of Residential Built Heritage in the Hot Summer and Cold Winter Zone — On the Example of a Traditional Courtyard House

赵沫沙, Hartwig M. Künzel, Schew-Ram Mehra
Zhao Mosha, Hartwig M. Künzel, Schew-Ram Mehra

摘要 我国夏热冬冷地区大量民居类建筑遗产的保护与再利用，是加强我国在城乡建设中历史文化保护传承的重要内容。针对我国夏热冬冷地区的民居类建筑遗产，大量研究已表明其夏季的室内物理环境偏离人居热舒适度限值较小，但是冬季偏离限值较大，阻碍了民居类建筑遗产的再利用。本文以夏热冬冷地区常见的合院建筑为例，通过现场调研、室内物理环境监测、能耗监测与模拟研究了有冬季采暖能耗的居住类建筑遗产的能耗现状。研究结果能提供夏热冬冷地区民居类合院建筑遗产冬季的室内用能行为与围护结构各部分的热量流失比例案例，促进形成民居类建筑遗产未来低碳利用的数据基础。

关键词 夏热冬冷地区；民居类建筑遗产；能耗状况

Abstract During the process of urban and rural construction, the conservation and use of residential heritages are of great importance in promoting the preservation and inheritance of China's history and culture. Existing studies have indicated that the deviation of the indoor thermal environment from the thermal comfort range in residential architectural heritages in the Hot Summer Cold Winter Zone is limited while that in winter is considerably high, preventing the reuse of residential heritages in this region. This article investigated the energy performance of a courtyard house with heating energy consumption in winter based on on-site survey, monitoring of indoor thermal environment and energy consumption, and numerical simulation. This study can provide the behavior of indoor energy consumption and the distribution of heat loss through building envelopes of residential courtyard heritage in winter in the Hot Summer Cold Winter zone, which forms the basis to promote the reuse of residential heritages in a low-carbon manner in the future.

Keywords Hot summer cold winter zone; Residential built heritages; Energy performance

赵沫沙，同济大学建筑与城市规划学院建筑系，助理教授。
Hartwig M. Künzel， 德国弗劳恩霍夫建筑物理研究所热湿研究部，教授。
Schew-Ram Mehra， 德国斯图加特大学声学与建筑物理研究所，教授。

相似性分析对单体建筑节能设计的机器学习准确度影响

Exploring the Impacts of Similarity Analysis on Machine Learning Models for the Energy Efficient Design of a Target Building

田志超，石邢
Tian Zhichao, Shi Xing

摘要 提升机器学习的预测准确度一直是研究者的目标。近年来，机器学习建筑能耗预测比赛结果表明数据预处理，如数据缩减，没有得到足够的重视。已有少量的研究采用了相似性分析这种数据缩减技术，然而其对机器学习模型的影响还不确定。本研究探索相似性分析对机器学习模型预测准确度的影响，具体是采用梯度提升机等算法预测能耗。结果表明尽管使用的数据量较少，在多数情况下，新方法构建的机器学习模型的预测准确度要高于传统方法构建的模型。相似性分析应成为数据驱动的单体建筑节能设计的数据预处理流程。

关键词 建筑节能设计；相似建筑；机器学习；单体建筑；数据驱动；集成学习

Abstract Improving the predictive accuracy of ML models is always the ambition of researchers. Recent building energy prediction competitions unveiled that preprocessing, especially data reduction, had not received enough attention. Previously, several studies have applied similarity analysis, a data reduction strategy, in building ML models. However, its impacts have not been explored exclusively. This study tends to unveil the unknown by conducting comparative studies of proposed and traditional methods. This study selects three machine learning algorithms, i.e., Random Forest, Gradient Boosting, and SVM to carry out the experiment. The results show that even with less data, in most cases, the ML models developed by the proposed method outperform ML models developed by traditional methods and in previous study. These findings establish that similarity analysis should be used as s general DDBEA process which also proves the value of small-size homogeneous data which can reduce the effects of data collection for DDBEA.

Keywords Building energy efficiency; Machine learning; Similarity analysis; Data-driven Ensemble learning

田志超，同济大学建筑与城市规划学院，博士后。
石邢，同济大学建筑与城市规划学院，教授，副院长。

通风及透明围护结构热工参数对夏热冬冷地区办公建筑室内热环境的影响

Influence of Ventilation and Transparent Envelope Thermal Parameters on the Indoor Thermal Environment of Office Buildings in Hot Summer and Cold Winter Zones

冉建东，郑焱红，李鹏南，冯雅，钟辉智，南艳丽
Ran Jiandong, Zheng Yanhong, Li Pengnan, Feng Ya, Zhong Huizhi, Nan Yanli

冉建东，重庆大学，博士研究生。
郑焱红，四川大学，硕士研究生。
李鹏南，四川师范大学，讲师。
冯雅，中国建筑西南设计研究院有限公司，教授级高级工程师。
钟辉智，中国建筑西南设计研究院有限公司，教授级高级工程师。
南艳丽，中国建筑绿色建造工程研究中心，高级工程师。

摘要 提高外窗的保温隔热性能是建筑节能领域研究的重点问题。但由于相关标准对室内外计算参数选择的要求，满足现有标准的外窗有可能并不适用于夏热冬冷地区的某些时间段。此外，外窗传热系数（K 值）的选择还与外窗的太阳得热系数（SHGC）及其是否通风有关，然而这些关系目前还未得到较为完整的研究。本文基于模拟的方法，研究了通风技术与透明围护结构热工设计两种被动式建筑设计方法结合使用时对室内热环境的影响。结果表明，通风是提高成都地区办公建筑室内热舒适的有效策略，而且最佳的透明围护结构的 SHGC 与 K 值应该与通风结合分析。

关键词 办公建筑；通风；太阳得热系数；传热系数

Abstract Improving the thermal insulation performance of windows is a key issue for research in the field of building energy efficiency. However, due to the requirements of the relevant standards for the selection of indoor and outdoor calculation parameters, it is possible that windows meeting the existing standards are not applicable to certain time periods in hot summer and cold winter regions. In addition, the choice of window K-value is also related to the solar heat gain coefficient of the window and whether it is ventilated or not, however, these relationships have not been studied more completely yet. This paper investigates the effects of two passive building design methods, ventilation technology and transparent envelope thermal design, on the indoor thermal environment when used in combination, based on a simulation approach. The results show that ventilation is an effective strategy to improve indoor thermal comfort in office buildings in Chengdu, and that the optimal SHGC and K values for transparent envelopes should be analyzed in conjunction with ventilation.

Keywords Office building; Ventilation; Solar heat gain coefficient; Heat transfer coefficient

街区形态对办公街区建筑能耗的协同影响规律及设计策略

Synergistic Influence of Block Morphology on Building Energy Consumption of Office Block and Energy-Saving Design Strategies

徐燊，李高梅 *，张海隆
Xu Shen, Li Gaomei, Zhang Hailong

徐燊，华中科技大学建筑与城市规划学院，湖北省城镇化工程技术研究中心，教授。
李高梅，华中科技大学建筑与城市规划学院，湖北省城镇化工程技术研究中心，博士研究生。
张海隆，中国铁路开发投资集团有限公司，工程师。

摘要 “双碳”战略目标导向下控制建筑领域的碳排放迫在眉睫，实现城市街区建筑能耗控制是关键环节。办公街区因其单位面积能耗强度大，成为重点关注的对象。本文以武汉市办公街区为例，评估了街区形态对建筑能耗的协同影响。首先，基于建筑类型学方法将办公街区分为 7 类，其次计算各街区的街区形态参数和建筑能耗强度；其次，采用统计学方法分析街区形态参数对建筑能耗的协同影响；最后，提出办公街区低能耗设计策略。研究结果显示：影响办公街区建筑能耗的关键形态参数组合为体型系数和容积率。

关键词 街区形态；办公街区；建筑能耗；协同影响；节能设计策略

Abstract With the dual carbon Goals-oriented, it is imperative to control carbon emissions in the building sector, and achieving control of energy consumption in urban blocks is a key aspect. Office blocks have become a key focus of research scholars due to their high energy intensity per unit area. In this paper, the aim is to evaluate the combined effect of office block morphology on building energy consumption taking office blocks in Wuhan as examples. First, the office blocks are divided into seven types based on an architectural typology approach. Then, the morphological parameters and building energy use intensity (EUI) for different blocks were calculated. Thirdly, the synergistic influence of block morphological parameters on building energy consumption was analyzed using statistical approach. Finally, based on the results, low-energy design strategies were proposed for office blocks. The results illustrated that the key morphological parameters synergistic influence the building energy consumption of office blocks are Building shape factor and Floor area ratio.

Keywords Block morphology; Office blocks; Building energy consumption; Synergistic influence; Energy-saving design strategies

屋面薄板架空遮阳的隔热降温性能理论评估

Theoretical Evaluation of the Thermal Isolation and Temperature Damping of Thin Sheet Overhead Sun-shading for Roof

冉茂宇，冉柯，王波
Ran Maoyu, Ran Ke, Wang Bo

摘要 薄板架空遮阳是炎热地区建筑外表隔热降温的被动式适宜技术，但目前尚缺乏其热性能全面评估的相关理论和方法。本文首先建立薄板架空遮阳下屋面综合温度计算模型，然后提出用附加热阻和附加热惰性指标来分别评估其隔热和降温性能，并推导给出了它们的计算式，最后基于夏季室外气象参数取值，评估了我国南方典型城市屋面薄板架空遮阳的热性能。研究结论为屋面薄板架空遮阳的热性能评估提供了科学依据和建议。

关键词 薄板架空遮阳；隔热降温；综合温度；附加热阻，附加热惰性

Abstract Thin sheet overhead sun-shading is the suitable passive technology for building thermal isolation and temperature damping in hot areas, but there is still a lack of relevant theories and methods to fully evaluate its thermal performance. This paper firstly established the air-solar temperature calculation model of roof with thin sheet overhead sun-shading. Then, additional thermal resistance and additional thermal inertia was respectively proposed to evaluate its thermal isolation and temperature damping performance，the calculation formula of them was also derived and given; Finally, based on the stipulated values of outdoor meteorological parameters in summer, the thermal performance of thin sheet overhead sun-shading in typical cities of southern China was evaluated. The studied results provide scientific basis and suggestions for the thermal performance evaluation of thin sheet overhead sun-shading for roof

Keywords Thin sheet overhead sun-shading; Thermal isolation and temperature damping; Air-solar temperature; Additional thermal resistance; Additional thermal inertia

冉茂宇，华侨大学建筑学院，教授，厦门市生态建筑营造重点实验室主任。
冉柯，华侨大学建筑学院，硕士研究生。
王波，深圳市绿创人居环境促进中心，助理工程师，节能审查负责人。

多智能体控制在建筑环境与能源应用中的研究综述

Research Review of Multi-agent Control in Building Environment and Energy Applications

王硕峰，李怀，孙育英
Wang Shuofeng, Li Huai, Sun Yuying

摘要 随着先进的绿色技术与系统的应用推广，暖通空调系统变化得更为复杂，多智能体控制作为人工智能先进的控制方法逐步得到重视。本文对暖通空调领域多智能体技术的国内外研究现状进行了分析总结。首先，本文概述了多智能体的基本概念、特点以及常用的优化算法；其次，综合分析了多智能体在建筑环境与能源领域的应用研究；最后，现有多智能控制体在应用中应注意的问题被指出。借助物联网等技术手段，多智能体控制也可以在智慧配电网领域中发挥重要作用。

关键词 多智能体；优化控制；能源系统；建筑节能；综述

Abstract With the application and popularization of advanced green technology and system, the changes of HVAC system are more complex, and multi-agent control as an advanced control method of artificial intelligence is gradually paid attention to. This paper analyzes and summarizes the research status of multi-agent technology in HVAC field at home and abroad. Firstly, this paper summarizes the basic concept, characteristics and common optimization algorithms of multi-agent. Secondly, the application research of multi-agent in the field of building environment and energy is analyzed comprehensively. Finally, the possible problems of existing multi-agent control are pointed out. With the help of technological means such as the Internet of Things, multi-agent control can also play an important role in the field of smart distribution networks.

Keywords Multi-agent system; Optimal control; Energy system; Building energy conservation; Overview

王硕峰，北京工业大学城建学部，硕士研究生。
李怀，中国建筑科学研究院，正高工程师。
孙育英，北京工业大学城建学部，副教授。

西北地区农村居住建筑能效提升支撑体系研究

Research on the Support System of Rural House Energy Efficiency Enhangement in Northwest China

豆琳，赵敏，王琼，赵敬源 *
Dou Lin, Zhao Min, Wang Qiong, Zhao Jingyuan

豆琳，长安大学，硕士研究生。
赵敏，长安大学，硕士研究生。
王琼，长安大学，高级工程师。
赵敬源，长安大学，教授。

摘要 西北地区处于严寒、寒冷和夏热冬冷地区，居民生活需求日益增加。为了有效提升西北地区农村居住建筑能效，本文通过资料收集和德尔菲法等方式，构建了西北地区建筑能效提升支撑体系，其主要为政策制度、技术、经济和能力四类二级指标体系，以专家评分均值≥3确定了三级近期最优指标和远期最优指标；采用 SPSS 26.0 软件统计分析德尔菲法问卷的可靠性，验证了支撑体系指标信效度。

关键词 西北地区；建筑能效；支撑体系；德尔菲法

Abstract The northwest region is located in severe cold region, cold region, hot summers and cold winters region, and the requirements in life of residents is increasing. In order to improve the energy efficiency of rural residential buildings in regions of Northwest China, this study used methods such as data collection, questionnaire surveys, the Delphi method, and expert discussions to construct a supporting system for enhancing building energy efficiency. The system consists of four main categories of secondary indicators: policy and regulations, technology, economy, and capacity. Through expert evaluations with an average score ≥ 3, the study identified the optimal short-term and long-term indicators at the third level. The reliability and validity of the indicators in the supporting system were verified using SPSS 26.0 software for statistical analysis.

Keywords Northwest China; Building energy consumption; Support system; Delphi

---

不同形式气凝胶墙体保温性能预测与对比

Prediction and Comparison of Thermal Insulation Performance of Different Aerogel-Based Walls

杨建明，吴会军 *，刘佳，张广鹏，庄浩杰
Yang Jianming, Wu Huijun*, Liu Jia, Zhang Guangpeng, Zhuang Haojie

杨建明，广东技术师范大学自动化学院，副教授。
吴会军，广州大学土木工程学院，研究员。
刘佳，广州大学土木工程学院，副教授。
张广鹏，广州大学土木工程学院，博士研究生。
庄浩杰，广东技术师范大学自动化学院，硕士研究生。

摘要 本文旨在探讨如何高效地应用气凝胶以实现建筑保温效果的最大化。通过建立气凝胶用于涂层、纤维板和混凝土三种墙体形式的保温性能理论模型，预测了气凝胶含量和密度对导热系数以及墙体保温性能的影响。在相同节能标准下，采用气凝胶纤维板的气凝胶用量最少，比涂层和混凝土形式分别减少 10%~14% 和 49%~54%。在基准墙体添加 20 毫米厚度气凝胶后，气凝胶纤维板的保温性能提升最高，达到 46.4%~53.8%。因此，宜优先采用气凝胶纤维板的方式，以减少气凝胶用量和提升经济性。

关键词 气凝胶；墙体；保温；节能；经济性

Abstract This article aims to explore how to efficiently apply aerogel to maximize building insulation performance. Theoretical models are established to predict the effects of aerogel content and density on thermal conductivity and wall insulation performance for three wall forms: coatings, fiberboards, and concrete. Under the same energy-saving standards, the use of aerogel fiberboards requires the least amount of aerogel, reducing the amount by 10%~14% and 49%~54% compared to coatings and concrete, respectively. After adding 20mm thickness of aerogel to the baseline wall, the insulation performance of aerogel fiberboards is the highest, reaching 46.4%~53.8%. Therefore, aerogel fiberboards is prioritized to reduce the amount of aerogel used and improve economic efficiency.

Keywords Aerogel; Wall; Insulation; Energy-saving; Economics

屋面薄板架空遮阳的隔热降温性能模态实测

Imitative Measurement of the Thermal Isolation and Temperature Damping of Thin Sheet Overhead Sun-shading for Roof

冉茂宇，王波，冉柯
Ran Maoyu, Wang Bo, Ran Ke

冉茂宇，华侨大学建筑学院，教授，厦门市生态建筑营造重点实验室主任。
王波，深圳市绿创人居环境促进中心，助理工程师，节能审查负责人。
冉柯，华侨大学建筑学院，硕士研究生。

摘要 薄板架空遮阳是炎热地区建筑外表隔热降温的被动式适宜技术。在前期研究中，作者曾就屋面薄板架空遮阳提出了综合温度计算模型，但尚未对该综合温度计算模型进行实验验证，同时也未实验考察风速和薄板太阳辐射吸收率对其热性能的影响。本文首先在人工气候房中搭建了试验小室，然后利用模拟灯实测了薄板架空遮阳对混凝土屋面的隔热降温效果，考察了薄板颜色和风速对其架空遮阳热性能的影响，最后就薄板温度的实测和理论计算进行了对比，间接验证了薄板架空遮阳下屋面综合温度计算模型的有效性。

关键词 屋面；薄板架空遮阳；隔热降温；综合温度；热工性能

Abstract Thin sheet overhead sun-shading is the suitable passive technology for building thermal isolation and temperature damping in hot areas. In previous studies, we proposed a calculation model of air-solar temperature for roof with thin sheet overhead sun-shading, but the air-solar temperature calculation model has not been experimentally verified, and the effects of wind speed and sheet color on the thermal performance have also not been investigated. In this paper, a test chamber with concrete roof was built in an artificial climate room, then the thermal isolation and temperature damping of thin sheet overhead sun-shading on the roof was measured by an imitative solar lamp, and the effects of wind speed and sheet color on the thermal isolation and temperature damping were investigated. Finally, the measured temperature of sheet is compared with the theoretical calculation, which indirectly verifies the validity of the air-solar temperature calculation model for roof with thin sheet overhead sun-shading.

Keywords Roof; Thin sheet overhead sun-shading; Thermal isolation and temperature damping; Air-solar temperature; Thermal performance

---

填料腔倾角对模块化热激活墙体冬季热性能与能耗的影响

The Effect of the Cavity's Inclination Angle on Thermal Performance and Energy Consumption of the Modular Thermally Activated Wall in Winter

陈萨如拉，陈天航，杨洋
Chen Sarula,
Chen Tianhang,
Yang Yang

陈萨如拉，安徽建筑大学建筑与规划学院，博士，讲师。
陈天航，安徽建筑大学建筑与规划学院，硕士研究生。
杨洋，合肥工业大学建筑与艺术学院，博士，讲师。

摘要 为探索寒冷地区典型冬季气象条件下模块化热激活复合墙体的动态热行为与节能潜力，本研究建立了复合墙体数值模型，基于敏感性分析方法研究了填料腔倾角度（θ 值）的影响。结果表明：将填料腔主轴偏向室内侧设置可获得更好的主动保温效果；随着 θ 值增加，复合墙体内表面热负荷与总注入热量分别呈现先减小后增大与先增大后减小的趋势；其他条件保持相同情况下，填料腔主轴向室内侧逆时针偏移 60° 时的复合墙体热性能综合最佳，而偏移角度为 0° 时的复合墙体的日累计一次能源消耗率可低至 1.94 kWh/m^2。

关键词 热激活建筑墙体；热特性；数值模拟；节能特性

Abstract In order to explore the dynamic thermal behavior and energy saving potential of the modular heat-activated composite wall under typical winter weather conditions in cold regions, a numerical model of the composite wall was established, and the influence of the inclination Angle (θ value) of the packing cavity was studied based on sensitivity analysis. The results show that better active insulation effect can be obtained by setting the main shaft of the packing chamber to the interior side. With the increase of θ value, the heat load and total injected heat on the inner surface of the composite wall decrease first and then increase, and then decrease, respectively. Under the same other conditions, the thermal performance of the composite wall is the best when the main axis of the packing chamber is offset to the interior side by 60° counterclockwise, while the cumulative daily primary energy consumption of the composite wall is as low as 1.94 kWh/m^2 when θ=0°.

Keywords Thermally activated Wall; Thermal characteristics; Numerical simulation; Energy-saving characteristics

夏热冬暖地区住宅围护结构隔热与辐射制冷材料实际应用效应对比分析

Comparative Analysis of the Actual Use of Heat Insulation and Radiation Cooling Materials for Residential Envelopes in Hot Summer and Warm Winter Regions

卜凡，燕达
Bu Fan, Yan Da

卜凡，清华大学建筑学院建筑节能研究中心，博士研究生。
燕达，清华大学建筑学院建筑节能研究中心，教授。

摘要 辐射制冷技术通过提高可见光波段辐射的反射率来降低建筑表面太阳辐射得热。然而，由于当外表面敷设有光谱制冷材料时，传热方向可能发生改变，这会导致隔热结构与辐射制冷无法协同工作。因此，需要进一步研究辐射制冷技术在夏热冬暖地区住宅建筑的应用潜力。本研究采用耦合模拟分析方法，以夏热冬暖地区住宅顶层为例，分别对比了隔热材料与辐射制冷材料的全年性能。

关键词 辐射制冷；住宅隔热；适应性分析；夏热冬暖地区

Abstract Radiative cooling technology reduces solar heat of gaining on building surfaces by increasing the reflectivity of radiation in the visible wavelengths. However, the direction of heat transfer may change when exterior surfaces are clad with spectrally cooled materials, which can result in insulation structures that do not work in concert with radiant cooling. Therefore, there is a need to further investigate the potential application of radiant cooling technology in residential buildings in hot summer and warm winter regions. In this study, a coupled simulation analysis was used to compare the year-round performance of insulation materials and radiant cooling materials separately for the top floors of residential buildings in hot summer and warm winter regions.

Keywords Radiative cooling; Residential insulation; Adaptability analysis; HSWW

第 2 篇

建筑与环境声学

城市公园声景初探——以重庆市花卉园为例

A Preliminary Study on the Soundscape of Urban Park — Taking Chongqing Flower Garden as an Example

李卓颖
Li Zhuoying

摘要 城市声景观是城市环境的重要组成部分，而城市公园作为城市的主要公共空间，起着调节生态、提供休闲场地的作用。想要提升城市公园整体环境品质，合理的声景设计是不可或缺的。本文结合文献资料和实地调研，初步探讨了我国城市公园声景研究现状，通过理论结合实际的方式，采用简化的声漫步法调查了一处城市公园声景，从自然声、人声、噪声三个方面结合实际分析其声景现状，最后提出高品质城市公园声景优化改造策略。

关键词 城市声景；公园声景观；城市公园；声漫步法

Abstract The urban soundscape is an important part of the urban environment. As the main public space of the city, urban parks play a role in regulating ecology and providing leisure venues. In order to improve the overall environmental quality of urban parks, reasonable soundscape design is indispensable. Based on literature and field research, this paper preliminarily discusses the research status of urban park soundscape in China. Through the combination of theory and practice, a simplified sound walk method was used to investigate the soundscape of an urban park. The current situation of the soundscape was analyzed from natural sound, human voice and noise. Finally, the optimization and transformation strategy of high-quality urban park soundscape was proposed.

Keywords Soundscape of the city; Soundscape of the park; City park; Sound walking method

李卓颖，重庆大学建筑城规学院，硕士研究生。

基于 CiteSpace 软件的声景观评价内容可视化分析

Visualization Analysis of Evaluation Content of Soundscape Based on CiteSpace Software

廉英奇，欧达毅 *
Lian Yingqi，Ou Dayi

摘要 声景观理论让人们重新认识了传统听觉行为，近年来的研究主要集中在声景观评价上。本文借助文献计量软件 CiteSpace 对近 10 年中文期刊中的声景观评价研究文献进行梳理，分析其研究热点及主题。结果显示，描述声景观主要使用声压级和声源类型，未来研究宜使用心理声学参数描述声景观；有些学者采用语义差异法评价整体声景观，后续研究者应提前筛选语义差异法所用词语；规范声景观研究方法，早日实现标准化声景研究。通过上述分析，明确声景观评价研究面临的问题，为我国未来声景观研究提供参考。

关键词 声景观；科学知识图谱；评价内容；声景元素

Abstract Soundscape makes people reconsider traditional auditory behavior. In recent years, research mainly focused on the evaluation of soundscape. With the help of bibliometric software, CiteSpace, this paper reviewed the literature of Chinese journals on the evaluation of soundscape in the past 10 years and analyzed their highlights and themes. The results showed that sound pressure level and sound source types were mainly used to describe the soundscape, and the psychoacoustic parameters should be used to describe the soundscape in future research. Some scholars used the semantic difference method to evaluate the overall soundscape, and subsequent research should select the words used in the semantic difference method in advance. Research methods of soundscape should be standardized as soon as possible. To sum up, the problems faced by the research of soundscape evaluation are clarified, and the paper can provide a reference for future soundscape research in China.

Keywords Soundscape; Mapping Knowledge Domains; Evaluation content; Soundscape elements

廉英奇，华侨大学建筑学院，博士研究生。
欧达毅，华侨大学建筑学院，教授，博士生导师。

疫情期间全球声环境的变化以及人对该变化的感知

Global Changes in the Acoustic Environment during the COVID-19 and Human Perception of These Changes

朱昕苗，刘芳芳 *，康健 *，武悦
Zhu Xinmiao, Liu Fangfang, Kang Jian, Wu Yue

朱昕苗，哈尔滨工业大学建筑学院，硕士研究生。
刘芳芳，哈尔滨工业大学建筑学院，副教授，博士生导师。
康健，英国伦敦大学学院巴特莱特建筑学院，教授；英国皇家工程院院士，欧洲科学院院士，中国工程院外籍院士。
武悦，哈尔滨工业大学建筑学院，副教授，博士生导师。

摘要 本研究系统地综述了疫情期间声环境产生的变化、人对该变化的感知，以及该变化对人生理和心理健康的影响。研究发现：①声级降低最大的是城市和人类活动声，达到了 30.5 dB(A)。在城市不同类型地区（活动、商业、交通、工业、住宅、安静、夜生活区）声级均降低，降低值在 3.3~8.8 dB(A)。②疫情以来监测到城市室外噪声级降低，人感知自然声更多并做出积极评价，而对室内声音的评价更消极。邻里噪声明显影响了人的烦恼度和压力水平。因此未来需要建立城市不同类型地区的数据库，以监测对人健康的声级阈值和临界点。

关键词 声环境；疫情；感知；生理健康；心理健康

Abstract The objective of this research is to systematically review the changes that occurred in the acoustic environment and the human perception of these changes and their physiological and mental health effects during the COVID-19 pandemic. These studies revealed: ① The greatest reduction in sound levels occurred in urban and anthropogenic sounds, reaching almost 30.5 dB(A). A trend of sound level reduction was reflected in different types of urban areas (active, commercial, traffic, industrial, residential, silence/quiet, and nightlife), with reductions ranging from 3.3 to 8.8 dB(A) in each area. ② As the sound level of urban outdoor noise reduces, the natural sound perceived by people is more dominant during the pandemic. People made positive comments on natural sounds, while for indoor sounds, people made more negative remarks. Neighborhood noise significantly affected people' s annoyance and stress levels. Database resources for different types of urban areas need to be created to monitor sound levels likely thresholds and tipping points for improved human health.

Keywords Acoustic environment; COVID-19; Perception; Physiological health; Mental health

---

开放式办公室吸声材料布置位置对声环境的影响研究

Study on the Influence of the Location of Sound Absorbing Materials on the Acoustic Environment in Open-plan Offices

陈楚希，欧达毅 *，张媛媛
Chen Chuxi, Ou Dayi, Zhang Yuanyuan

陈楚希，华侨大学建筑学院，硕士研究生。
欧达毅，华侨大学建筑学院，教授，博士生导师。
张媛媛，华侨大学建筑学院，博士研究生。

摘要 开放式办公室的声环境对办公人员的舒适度和工作效率有重要影响。在开放式办公室中布置吸声材料能有效改善室内的声环境。本文基于 Odeon 建筑声学模拟软件，建立了开放式办公室声环境模拟模型，并通过实地测量验证了该模型的有效性。基于该模型，本文对开放式办公室吸声材料布置位置进行研究，分析了在开放式办公室的天花板、地板和墙面三种位置设置吸声材料的声环境特点。本文所做研究可以为开放式办公室声环境的理论研究和优化设计提供参考。

关键词 建筑声学；声环境；开放式办公室；吸声材料；Odeon

Abstract The acoustic environment of open-plan offices has an important impact on the comfort and efficiency of staff. Placing sound-absorbing materials in open-plan offices can effectively improve the indoor acoustic environment. Based on Odeon architectural acoustics simulation software, this paper establishes an open-plan office acoustic environment simulation model, and verifies the effectiveness of the model through field measurements. Based on this model, this article studies the placement of sound-absorbing materials in open offices and analyzes the acoustic environment characteristics of sound-absorbing materials installed in three positions: ceiling, floor, and wall. The research conducted in this article can provide a reference for theoretical research and optimized design of open-plan office acoustic environment.

Keywords Architectural acoustics; Acoustic environment; Open-plan offices; Sound-absorbing materials; Odeon

小尺度开敞空间声音围合度研究

Small-scale Open Space Sound Enclosure Established

马建军，毛琳箐，张鑫
Ma Jianjun, Mao Linqing, Zhang Xin

马建军，石家庄铁道大学建筑与艺术学院，硕士研究生。
毛琳箐，石家庄铁道大学建筑与艺术学院，副教授，副院长。
张鑫，石家庄铁道大学建筑与艺术学院，硕士研究生。

摘要 小尺度开敞空间是城市景观构成中不可或缺的一部分，对其进行合理利用，有助于提升空间品质。本文结合了 Odeon 声场仿真模拟与现场声环境测试，通过对所研究的小尺度开敞空间进行分类，并结合主观问卷调查，以期建立主观与客观融合的声音围合度概念，为城市景观构建与设计提供参考。

关键词 小尺度；开敞空间；声音围合度；声环境

Abstract Small-scale open spaces are an integral part of the urban landscape, and their rational use can help improve the quality of the space. This paper combines the Odeon sound field simulation and on-site acoustic environment test, and classifies the studied small-scale open space, in order to establish a classification method for small-scale open space under the concept of the sound enclosure, and provide a reference for urban landscape construction and design.

Keywords Small Scale; Open Space; Sound enclosure; Sound environment

基于几何声学模拟下语言清晰度 C50 理论式适用性分析

Research on the Applicability of the C50 Theoretical Formula for Speech Intelligibility Based on Geometric Acoustic Simulation

陈科吉，何钰昆，初楚，张梦晗
Chen Keji, He Yukun, Chu Chu, Zhang Menghan

陈科吉，东京大学大学院工学系研究科建筑学专攻，博士研究生。
何钰昆，京都大学大学院工学研究科建筑学专攻，博士研究生。
初楚，清华大学建筑学院，助理研究员，工学博士。
张梦晗，东京大学物性研究所，硕士研究生。

摘要 营造良好的声环境需要仔细考虑空间的吸声性能，然而目前对于日常空间的吸声性能并没有详细的设计标准。确定吸声性能和语言传播性能之间的关系是制定日常空间的吸声设计准则的一个关键。在扩散声场的假设下，存在语言清晰度 C50 的理论公式，然而它对各种房间条件的适用性尚不明确。在本研究中，基于几何声学模拟进行案例研究，以房间容积、声源－受声点分布与距离、空间吸声分布、平均吸声系数为参数，以此获得各个条件下语言清晰度 C50 的模拟值，并分析理论值和模拟值之间对应关系，研究在各个条件下的适用性。

关键词 几何声学模拟；语言清晰度；C50；理论值；模拟值；适用性

Abstract Creating a good acoustic environment requires deep consideration of a space's sound absorption performance. However, there are currently no detailed design standards for the sound absorption performance of normal spaces. Establishing the relationship between sound absorption performance and speech transmission performance is a key factor in developing design guidelines for sound absorption of normal space. Under the assumption of a diffuse sound field, there exists a theoretical formula for speech clarity C50, but its applicability to various room conditions is not yet clear. In this study, a case study was conducted based on geometric acoustic simulation. Room volume, sound source-receiver distribution and distance, spatial sound absorption distribution, and average sound absorption coefficient were used as parameters to obtain the simulation values of speech clarity C50 under various conditions. The corresponding relationship between theoretical and simulated values was analyzed to investigate their applicability under various conditions.

Keywords Geometric acoustic simulation; Speech intelligibility; C50; Theoretical values; Simulated values; Applicability

倾斜建筑立面与垂直建筑立面对周围声环境影响对比研究

A Comparative Study on the Impact of Inclined and Vertical Building Facades on the Surrounding Sound Environment

钱君如，曾庆勇，袁童谣，李颖，李争光 *
Qian Junru, Zeng Qingyong, Yuan Tongyao, Li Ying, Li Zhengguang

钱君如，曾庆勇，袁童谣，李颖，浙江科技学院建筑系，本科生。
李争光，浙江科技学院建筑系，讲师。

摘要 倾斜建筑立面在城市中已屡见不鲜，其对周围声环境的影响有其特殊性。本研究分别选取了周围声环境相似的一处倾斜建筑立面和垂直建筑立面，通过现场声学测量、声学模拟和噪声烦恼度现场问卷调查，比较了 2 种建筑立面对周围声环境影响的差异。结果表明，向道路噪声源一侧倾斜约 50° 的建筑立面附近实测噪声比垂直建筑立面高 1.1 dB，声学模拟结果则高 1.6 dB；倾斜建筑立面附近噪声烦恼度比垂直建筑立面附近要高，原因可能是 2 处噪声级不同、倾斜建筑立面处存在视听不一致现象，人群性别和噪声敏感性对噪声烦恼度有显著影响。

关键词 建筑立面；噪声烦恼度；视听不一致；现场调查

Abstract Nowadays, inclined building facades are commonplace in cities. Its impact on sound propagation paths and the subjective perception of community noise is unique. An inclined building facade and a vertical building facade with a similar surrounding acoustic environment were selected in this study to compare the impact of the two building facades on the surrounding acoustic environment through on-site acoustic measurement, acoustic simulation and on-site questionnaire survey of noise annoyance. The results show that the measured noise level near the inclined building facade with an inclination angle of about 50° to the road noise source side is 1.1dB higher than that of the vertical building facade, while the acoustic simulation result is 1.6dB higher. Noise annoyance near the inclined building facade is higher than that of the vertical building façade, which may be due to the difference in noise levels between the two locations, or the presence of audio-visual inconsistency at the inclined building facade. People' s gender and noise sensitivity have a significant impact on noise annoyance.

Keywords Building facade; Noise annoyance; Audio-visual inconsistency; Site investigation

---

大运河遗产廊道历史声景观史料研究——以沧州段为例

Study of the Historical Soundscape Along the Grand Canal Heritage Corridor in the Cangzhou

张鑫，毛琳箐，马建军
Zhang Xin, Mao Linqing, Ma Jianjun

张鑫，石家庄铁道大学建筑与艺术学院，硕士研究生。
毛琳箐，石家庄铁道大学建筑与艺术学院，副教授，副院长。
马建军，石家庄铁道大学建筑与艺术学院，硕士研究生。

摘要 大运河遗产廊道的构建为沿途生态系统、文化遗产等构成要素的保护提供了整体性、活态化空间，其中声音要素作为活化的关键之一。本研究以大运河遗产廊道沧州段为例，旨在通过史料搜集，整理关于声景相关描述，为活态化保护提供相应的素材。

关键词 大运河；遗产廊道；历史声景；声环境

Abstract The construction of the Grand Canal heritage corridor provides a holistic living space for the protection of the ecological system, cultural heritage and other elements along the way, of which the sound element is one of the keys to activation. Taking the Cangzhou section of the heritage corridor of the Grand Canal as an example, this study aims to collate relevant descriptions of soundscape through the collection of historical materials, and provide corresponding materials for living protection.

Keywords Grand Canal; Heritage corridor; Historic soundscape; Sound environment

国家标准《通风消声器》产品性能指标解读和思考

Interpretation and View on Product Technical Specification of Chinese National Standard *Ventilation silencer*

闫国军，方庆川，徐春
Yan Guojun, Fang Qingchuan, Xu Chun

闫国军，中国建筑科学研究院有限公司，正高级工程师，声学室主任。
方庆川，深圳中雅机电实业有限公司，教授级高级工程师。
徐春，中国建筑科学研究院有限公司，教授级高级工程师。

摘要 产品国家标准《通风消声器》（GB/T 41318—2022）是首部针对通风空调系统工程中配套使用的消声器而制定的产品国家标准，于2022年3月发布，2022年10月起开始实施。本文从标准的编制背景、原则和编制思路出发，介绍本标准的主要内容，重点介绍标准编制过程中，产品性能指标要求的选择原则和指标限值确定依据，以及标准实施中需要注意的问题，供标准使用者参考。

关键词 通风消声器；标准；技术指标；通风空调系统

Abstract *Ventilation silencer*(GB/T 41318—2022) is the first product national standard for silencers in ventilation and air conditioning systems engineering, and it was issued in March 2022 and has been implemented since October 2022. Based on the compilation background, principles and ideas, this paper systematically introduces the main technical contents of the *Ventilation silencer*, focusing on the selection principle of the main technical requirement and the basis for determining the requirement indicators limit, and the problems needing attention in the implementation of the standard, for reference of standard users.

Keywords Ventilation silencer; Standard; Technical specification; Ventilation and air conditioning system

---

国家标准《通风消声器》产品性能试验方法解读

Interpretation on Product Performance Test Method of Chinese National Standard *Ventilation silencer*

闫国军，徐春，赵丹，刘庆，徐立军
Yan Guojun, Xu Chun, Zhao Dan, Liu Qing, Xu Lijun

闫国军，中国建筑科学研究院有限公司，正高级工程师，声学室主任。
徐春，中国建筑科学研究院有限公司，教授级高级工程师。
赵丹，中国建筑科学研究院有限公司，高级工程师。
刘庆，中国建筑科学研究院有限公司，高级工程师。
徐立军，中国建筑科学研究院有限公司，助理工程师。

摘要 国家产品标准《通风消声器》（GB/T 41318—2022）是首部针对通风与空调系统工程中配套使用的消声器而制定的产品国家标准，于2022年3月发布，2022年10月起开始实施。本文重点介绍标准中各类产品性能指标对应试验方法和计算方法。

关键词 通风消声器；A计权插入损失；全压损失；单位面积气流噪声A计权声功率级；漏风量；耐高温

Abstract The national product standard *Ventilation silencer* (GB/T 41318—2022) was issued in March 2022 and has been implemented since October 2022. Based on the compilation background, principles and ideas, this paper mainly introduces the test methods and calculation methods of main product performance indicators in the standard.

Keywords Ventilation silencer; A-weighted insertion loss; A-weighted sound power level of flow noise per unit face area; Total pressure loss coefficient; Air leakage; High-temperature resistance

住宅建筑全过程噪声控制实现路径的思考

View on Implementation Path of the Whole-Process Noise Control of Residential Building

闫国军，陈钊贤，赵启元，刘青云，周勇
Yan Guojun, Chen Zhaoxian, Zhao Qiyuan, Liu Qingyun, Zhou Yong

闫国军，中国建筑科学研究院有限公司，正高级工程师，声学室主任。
陈钊贤，中国建筑科学研究院有限公司，硕士研究生。
赵启元，中国建筑科学研究院有限公司，工程师。
刘青云，中国建筑科学研究院有限公司，助理工程师。
周勇，中国建筑科学研究院有限公司，高级工程师。

摘要 随着《中华人民共和国噪声污染防治法》和《"十四五"噪声污染防治行动计划》的实施，社会对噪声污染的防治越来越重视。住宅作为人居住的场所，室内声环境水平是否宁静会对人产生重要影响。本文通过总结近年来10多个住宅项目全过程噪声控制实践经验，对住宅建筑全过程噪声控制的实施步骤和方案进行介绍，并对今后提高住宅声环境水平提出了一些想法和思考。

关键词 住宅建筑；噪声控制；全过程；宁静住宅评价

Abstract With the implementation of the *Law of the People's Republic of China on Noise Pollution Prevention* and *Control* and *the 14th Five-Year Plan for Noise Pollution Prevention and Control*, more and more attention has been paid to the prevention and control of noise pollution. The residential building is the place where people live, and it will have an important impact on whether the indoor acoustic environment is quiet or not. This paper introduces the implementation steps and schemes of the whole-process noise control of residential buildings by summarizing the practical experience of more than 10 residential building projects in recent years, and putting forward some ideas and thoughts on improving the acoustic environment level of residential buildings in the future.

Keywords Residential building; Noise control; Whole-process; Assessment of quiet residential building

---

寒地城市公共开放空间冬季声景现状及改进研究
——以哈尔滨为例

Research on the Current Situation and Improvement of Winter Soundscape in Urban Open Public Spaces of Cold Regions Cities—Taking Harbin as an Example

唐征征，韩振坤，吕海景，张卷舒
Tang Zhengzheng, Han Zhenkun, L ü Haijing, Zhang Juanshu

唐征征，哈尔滨工业大学建筑学院，讲师。
韩振坤，哈尔滨工业大学建筑学院，副教授。
吕海景，东北农业大学艺术学院，副教授，系主任。
张卷舒，哈尔滨工业大学建筑学院，副教授。

摘要 城市公共开放空间是承载城市活力的重要空间载体，目前寒地城市公共开放空间冬季因地理位置和寒冷气候等原因，以及人们从事户外活动和公共交往特点，在一定程度上决定了人们在城市开放空间时间的集中性和人群的聚集性，需要提升公共开放空间声景质量，从而满足居民冬季日常生活休闲和社会交往的需求。本文基于此问题，以哈尔滨为例，通过对哈尔滨公共开放空间冬季声源的种类和声景现状调研与分析，探寻寒地城市公共开放空间冬季声景现状存在的主要问题，并据此提出了一些改善的建议，以期为增强寒地城市冬季公共开放空间的吸引力和活力提供一定依据和策略。

关键词 寒地城市；冬季声景；声源种类

Abstract Urban open public space is an important spatial carrier that carries urban vitality. Currently, due to geographical location and cold climate, as well as the characteristics of outdoor activities and public communication, people's concentration of time and crowd aggregation in urban open spaces are determined to a certain extent in winter. It is necessary for us to improve the soundscape quality of open public spaces. In order to meet the needs of residents for daily leisure and social interaction in winter. Based on this issue, this article takes Harbin as an example to investigate and analyze the types of winter sound sources and the current situation of the soundscape in open public spaces in Harbin. It explores the main problems in the current situation of winter soundscape in cold cities' open public spaces, and proposes some improvement suggestions based on this, in order to provide a certain basis and strategy for enhancing the attractiveness and vitality of winter open public spaces in cold cities.

Keywords Cold cities; Winter soundscape; Type of sound source

空间形态对不同建筑布局的居住小区交通噪声水平的影响——以成都市为例

The Influence of Spatial Morphological on the Traffic Noise Level of Residential Quarters with Different Building Layouts—Take Chengdu as a Case

强银平，于佳佳，吴聃，张樱子

Qiang Yinping, Yu Jiajia, Wu Dan, Zhang Yingzi

强银平，西南交通大学建筑学院，硕士研究生。
于佳佳，四川省建筑科学研究院有限公司，高级工程师。
吴聃，西南交通大学建筑学院，讲师。
张樱子，西南交通大学建筑学院，教授，学院助理、建筑系系主任。

摘要 大量研究证明，建筑布局是影响居住小区室外声环境的重要因素，但在不同的建筑布局方式中，空间形态参数对居住小区空间交通噪声水平的影响还未探明。本文根据三种主要的建筑布局方式，各选择 30 个居住小区，利用 Cadna/A 模拟居住小区的空间交通噪声分布。选择并分析用地面积、建筑密度、绿地率、容积率、平均建筑高度、相邻建筑的平均间距、第一排建筑与噪声源的距离以及沿街开口总宽度八种空间形态参数与居住小区空间噪声分布的关系。结果表明，居住小区的建筑布局方式不同，空间形态参数会对小区空间交通噪声水平产生不同的影响。

关键词 建筑布局；空间形态参数；Cadna/A；噪声水平

Abstract A large number of studies have shown that building layout is an important factor affecting the outdoor acoustic environment of residential quarters. However, in different building layouts, the influence of spatial morphological parameters on the spatial traffic noise level of residential quarters remains to be studied. In this paper, 30 residential quarters are selected according to three typical building layouts, and Cadna/A is used to simulate the spatial traffic noise distribution of residential quarters. Select and analyze the relationship between spatial morphological parameters including land area, building density, greening rate, floor area ratio, average building height, average spacing of adjacent buildings, distance between the first row of buildings and noise sources, total width of openings along the street and spatial noise level in residential quarters. The results show that the spatial morphological parameters affecting the spatial traffic noise level of residential quarters are also different due to the different building layouts of residential quarters.

Keywords Building layout; Spatial morphological parameters; Cadna/A; Noise level

---

超大空间混响时间测量方法对比研究

Comparison of Reverberation Time Measurement Methods in Extra-Large Room

赵宇辰，徐漪荃，岳文瑄，闵鹤群

Zhao Yuchen, Xu Yiquan, Yue Wenxuan, Min Hequn

赵宇辰，东南大学建筑学院，博士研究生。
徐漪荃，东南大学建筑学院，博士研究生。
岳文瑄，东南大学建筑学院，硕士研究生。
闵鹤群，东南大学建筑学院，副教授，硕士、博士研究生导师。

摘要 超大空间混响时间的测量方法在国内外标准中尚无明确规定。应用脉冲反向积分法与中断声源法对容积约为 1.8 万 m^3 和 23 万 m^3 的超大封闭空间内的多个测点进行混响时间测量，并与 Raynoise 仿真模拟结果相对比。研究表明，脉冲反向积分法的可重复性好于中断声源法，多次测量的标准误差较小。脉冲反向积分法在低频的测量结果低于中断声源法，在高频的测量结果则偏高，但这一误差不超过 10%。只要采用标准化的测量流程，两种方法得到的超大空间混响时间的测量结果都是可靠的。

关键词 超大空间；混响时间；脉冲反向积分法；中断声源法；现场测量

Abstract The measurement method of reverberation time in a huge space is not specified in the existing standards. The reverberation time of two enclosed rooms with a volume of 18000 m^3 and 230000 m^3 was measured by using the pulse reverse integration method and the interrupted sound source method, and compared with the results of simulation results obtained by Raynoise. The results show that the repeatability of the pulse reverse integration method is better than that of the interrupted sound source method, and the standard error of multiple measurements is smaller. The measurement results of the pulse reverse integration method are smaller at low frequencies and higher at high frequencies, but this difference does not exceed 10%. The reverberation time in extra-large room obtained by the two measurement methods is reliable when applied with standardized measurement process.

Keywords Extra-large room; Reverberation time; Pulse reverse integration method; Interrupted sound source method; Measurement

丰台站大型候车空间的建筑声学设计

Architectural Acoustic Design of Large Waiting Space in Fengtai Station

胡文林，何宾，马辉，刘冀钊
Hu Wenlin, He Bin,
Ma Hui, Liu Jizhao

胡文林，中国铁路设计集团有限公司城市轨道交通数字化建设与测评技术国家工程研究中心，高级工程师，分室副主任。
何宾，中国铁路设计集团有限公司城市轨道交通数字化建设与测评技术国家工程研究中心，高级工程师。
马辉，中国铁路设计集团有限公司建筑设计研究院，高级工程师。
刘冀钊，中国铁路设计集团有限公司城市轨道交通数字化建设与测评技术国家工程研究中心，正高级工程师，分室主任。

摘要　北京铁路枢纽丰台站拥有三个大型候车空间，分别为高架候车厅、地下快速进站厅和进站集散厅，需要开展建筑声学设计。本文较详细地剖析不同候车空间的主要声源构成和建筑声学设计目标，分析了空间几何特征对吸声材料作用效果的影响差异，针对建筑室内装饰风格研究提出合理可行的声学措施，并对部分区域的建筑声学指标实际进行了现场实测。

关键词　大型候车空间；建筑声学；铁路枢纽；丰台站；混响时间

Abstract　Fengtai Station, the Beijing railway hub, has three large waiting spaces, namely, the elevated waiting hall, the underground rapid entrance hall and the entrance and distribution hall. Architectural acoustic design is required. This paper analyzes in detail the main sound source composition and architectural acoustic design objectives of different waiting spaces, analyzes the difference in the influence of space geometric characteristics on the effect of sound absorption materials, and puts forward reasonable and feasible acoustic measures for the study of architectural interior decoration style. Part of architectural acoustic indicators in some areas are measured.

Keywords　Large waiting spaces; Architectural acoustic design; Railway hub; Fengtai Station; Reverberation time

---

家庭办公室的视听交互：植物和交通噪声对阅读的影响

Visual-Auditory Interaction in Home Offices: The Impact of Indoor Plants and Traffic Noise on English Reading Comprehension

张媛媛，欧达毅 *，陈秋，陈楚希，曲冠华
Zhang Yuanyuan, Ou Dayi,
Chen Qiu, Chen Chuxi,
Qu Guanhua

张媛媛，华侨大学建筑学院，博士研究生。
欧达毅，华侨大学建筑学院，教授，博士生导师。
陈秋，温州市鹿城区土地储备出让中心，科员。
陈楚希，华侨大学建筑学院，硕士研究生。
曲冠华，天津大学建筑学院，博士研究生。

摘要　本研究以 22 名中国大学生为对象，探究了居家办公环境中，室内植物（有或无）和不同等级（35、45、50、55 和 60 dBA）的交通噪声对英语阅读理解任务的影响。结果显示，在有植物的环境下，受试者的眼动生理指标（如平均注视时间等）表现更佳，且在 50 dBA 的噪声水平下，有无植物的影响最大。因此，研究表明室内植物对进行远程工作的个体有益。本研究从生理的角度强调了家庭办公环境中视听交互的重要性。

关键词　室内植物；交通噪声水平；眼动生理指标；英语阅读理解；家庭办公室

Abstract　This study investigated the impact of the presence or absence of indoor plants and varying levels of traffic noise (35, 45, 50, 55, and 60 dBA) on English reading comprehension task (ERCT) in a remote work setting, utilizing a cohort of 22 Chinese university students. It was found that the presence of indoor plants resulted in superior eye movement physiological indicators, such as increased average fixation duration. The most notable difference between having plants and not having plants was observed at a noise level of 50 dBA. Thus, it was indicated by the study that indoor plants could potentially benefit individuals undertaking remote work. The importance of audio-visual interaction within home office settings was underscored by this physiological investigation.

Keywords　Indoor plants; Traffic noise level (TNL); Eye movement physiological indicators; English reading comprehension task (ERCT); Home offices

剧院建筑声环境实测与模拟研究

Measurement and Simulation of Acoustic Environment in Theater Buildings

徐涵，欧达毅 *
Xu Han, Ou Dayi

摘要 本文以闽南某剧院为例，首先，开展声环境参数实地测试，包括并通过构建实地舞台空间网格模数，开展脉冲响应测量。其次，运用 Odeon 软件进行声场建模，并建立剧院模型，并与实测结果对比，进行声场仿真模拟验证。最后，通过实测和模拟结果，综合探讨了剧院早期衰变时间、混响时间、音乐明晰度、语言清晰度等声学参量与发声位置的关系。研究发现：①发声位置对混响时间、早期衰变时间、音乐明晰度、语言清晰度有明显影响；②侧轴声源的 EDT、T30 高于中轴声源；③在 0~12 m 舞台进深区间，随着进深距离增加，EDT、T30、C80、D50 随之升高。

关键词 剧院；建筑声学参量；发声位置；倍频程；仿真模拟

Abstract Taking a theater in southern Fujian as an example, this paper firstly conducts field tests of acoustic environment parameters, including building a spatial grid model of the live dance hall, and carrying out impulse response measurements. Secondly, use Odeon software to enter the sound field model, build a theater model, and compare it with the actual measurement results to enter the sound field simulation test. Finally, through the actual measurement and simulation results, the relationship between the acoustic parameters such as the early evolution time of the theater, reverberation time, music brightness, language clarity and sound distance is comprehensively discussed. Research findings: ① The voice position has a significant impact on reverberation time, early decay time, music clarity, and speech clarity; ② The EDT and T30 of the side axis sound source are better than the central axis sound source; ③ When entering the deep area of the 0 ~12 m stage, as the depth of entry increases, the EDT, T30, C80, and D50 increase accordingly.

Keywords Theater; Architectural acoustic parameters; Voice position; Octave; Simulation

徐涵，华侨大学建筑学院，硕士研究生。
欧达毅，华侨大学建筑学院，教授，博士生导师。

开放式办公室内个体差异对声环境感知和工作效率的影响

The Impacts of Individual Differences on Acoustic Environmental Perceptions and Work Performance in Open-plan Offices

康升娴 *，麦卓明，欧达毅，周新欣
Kang Shengxian, Mak Cheuk Ming, Ou Dayi, Zhou Xinxin

摘要 随着开放式办公室的普及，其室内声环境缺陷所引起的员工对办公环境的不满，愈发受到研究学者的重视。糟糕的开放式办公室声环境会降低员工的工作效率、工作和环境满意度。噪声类型、室内声压级大小、语言清晰度都是影响员工环境感知的重要声学因素。此外，员工间的个体差异也会影响到员工的声环境感知继而引起工作效率的改变。本研究参考国际标准 ISO 3382-3:2022 中的声环境质量划分标准，利用声学仿真模拟建立 4 个声环境质量迥异的开放式办公室声场，并在实验室中利用可听化技术探讨性别、教育背景、噪声敏感度等个体差异因子对员工声环境感知和工作效率的影响。

关键词 开放式办公室；声环境感知；个体差异；工作效率

Abstract With the popularity of open-plan offices, the defects of the indoor acoustic environment have resulted in occupants' environmental dissatisfaction and have attracted researchers' attention. Poor acoustic quality in open-plan offices can reduce work performance and environmental satisfaction. Noise type, indoor sound pressure level, and speech intelligibility are the vital acoustic factors that affect occupants' environmental perception. Individual differences also affect occupants' perception of the acoustic environment. In this study, according to the acoustic quality classification method recommended by ISO 3382-3:2022, four open office models with different acoustic qualities were built using acoustic simulation software. Subsequently, auralization technology was used in the laboratory to explore the effects of individual differences (e.g., gender, educational background, and noise sensitivity) on acoustic environmental perceptions and work performance.

Keywords Open-plan offices; Acoustic environmental perception; Individual difference; Work performance

康升娴，湖南科技大学建筑与艺术设计学院，博士研究生。
麦卓明，香港理工大学建设及环境学院，教授，博士生导师。
欧达毅，华侨大学建筑学院，教授，博士生导师。
周新欣，广州市设计院集团有限公司，中级工程师。

声源朝向对教室语言清晰度的影响

Effect of Speaker Orientation on Speech Intelligibility in Classrooms

梁林达*，任苗，叶朝铭
Liang Linda, Ren Miao, Ye Chaoming

摘要 本文利用语言传输指数（STI）探究声源朝向对教室语言清晰度的影响。选取三个不同尺寸的教室，采用仿真嘴作为声源，在不同声源朝向下测得不同窗帘状态和接收点的房间脉冲响应，并计算相应的 STI。结果表明，对于距离较近的接收点而言，声源朝向对 STI 的影响较大。相比于中型和大型教室，小型教室的 STI 明显要大得多。不同声源朝向下的 STI 变化幅度基本不大于 0.08，远小于无反射环境下仅由声源指向性导致的 STI 变化范围。窗帘的吸声使得不同声源朝向下 STI 的变化范围整体上增大。本文研究结果对于改善教室内听闻环境具有重要的指导意义。

关键词 语言清晰度；教室；声源朝向；语言传输指数

Abstract This study employs the speech transmission index (STI) to investigate the influence of speaker orientation on speech intelligibility in classrooms. Three classrooms of different sizes were selected, with an artificial mouth used as the speaker to measure the room impulse responses at different receiving points and curtain conditions under varying speaker orientations, which were then used to calculate corresponding STI values. The results indicate that for closer receiver points, the influence of speaker orientation on STI is relatively small. Compared to medium-sized and large classrooms, the STI of small classrooms is significantly larger. The STI variation range under different speaker orientations is basically not greater than 0.08, which is much smaller than those caused solely by the speaker's directivity in a non-reflective environment. The sound absorption provided by the curtains increases the STI variation range under different speaker orientations. The findings of this study provide important guidance for improving the listening environment in the classroom.

Keywords Speech intelligibility; Classroom; Speaker orientation; Speech transmission index

梁林达，广西大学土木建筑工程学院 助理教授。
任苗，广西大学土木建筑工程学院，硕士研究生。
叶朝铭，广西大学土木建筑工程学院 本科生。

人为噪声骤降下的城市住宅声环境

Sound Environment in an Apartment Building Under Significant Anthropogenic Noise Reduction

杨婷婷，康健，陈易
Yang Tingting, Kang Jian, Chen Yi

摘要 宁静小区的建设近年来受到越来越多的关注。2022 年春季上海市采取静态管理以控制新冠病毒流行病传播。本研究分别在静态和常态管理两个时间段对上海一栋 14 层的住宅楼进行噪声监测，该楼由其他建筑物所包围和屏蔽。结果显示，静态管理期间的等效声级的平均值低于常态管理期，昼间和夜间分别相差 3.6 dB 和 3.1 dB。在噪声建筑高度分布方面，两个时间段基本相似，均呈低层“低”、高层“高”和中层“差异大”的模式。宁静小区的住宅设计应考虑住宅的建筑高度、周边绿化及昼夜等因素，以促进声与健康的营造与融合。

关键词 宁静小区；人为噪声；声环境；城市安静区

Abstract The construction and design of quiet communities as an approach to maintain a harmonious and tranquil urban living environment draws increasing attention in recent years. The city of Shanghai experienced a city-wide lockdown to control the spread of COVID-19 in 2022, presenting conditions of quietness. Here, we conducted noise monitoring in a 14-story apartment building that was surrounded and shielded by other buildings in a typical urban community during and after the lockdown. The mean value of equivalent continuous A-weighted sound pressure level of all 14 floors during lockdown was lower than that after lockdown, with level differences of 3.6 and 3.1 dB during the daytime and night-time periods, respectively. In terms of noise distribution of building heights, it is basically similar during and after lockdown, with a pattern of "low" low-rise, "high" high-rise, and "high variance" mid-rise. We suggest that in practice, it is useful to consider the specific floor or time of day to promote the creation and integration of sound and health.

Keywords Quiet community; Anthropogenic noise; Sound environment; Urban quiet area

杨婷婷，同济大学建筑与城市规划学院建筑系，助理教授。
康健，英国伦敦大学学院巴特莱特建筑学部，教授。
陈易，同济大学建筑与城市规划学院建筑系，教授。

某报告厅建筑声学设计与实测分析

Architectural Acoustics Design and Field Measurement Analysis in a Lecture Hall

刘青云，闫国军，赵启元，吴伟斌，陈钊贤
Liu Qingyun, Yan Guojun, Zhao Qiyuan, Wu Weibin, Chen Zhaoxian

刘青云，中国建筑科学研究院有限公司，助理工程师。
闫国军，中国建筑科学研究院有限公司，正高级工程师，声学室主任。
赵启元，中国建筑科学研究院有限公司，工程师。
吴伟斌，中国建筑科学研究院有限公司，高级工程师。
陈钊贤，中国建筑科学研究院有限公司，硕士研究生。

摘要 某报告厅观众厅容积约为 7500 m^3，共 859 座，以会议功能为主。音质设计主要应控制合适混响时间，保证使用扩声系统时语言清晰。根据观众厅规模及使用功能，观众厅中频 500~1000 Hz 满场混响时间宜为 1.2s±0.1 s。采取措施包括体型调整、观众厅通过织物软椅及部分墙面吸声构造，估算 500 Hz 及 1000 Hz 满场混响时间 1.12 s、1.29 s。在未安装座椅但室内装修基本完成时报告厅空场混响时间测试评估，并通过赛宾公式估算其 500 Hz 及 1000 Hz 满场混响时间平均值为 1.20 s、1.30 s，达到预期设计指标。

关键词 报告厅；建筑声学；混响时间；实测

Abstract The net volume of the lecture hall is about 7500 m^3, and the audience capacity is about 859 seats, mainly for conference functions. The architectural acoustic design should mainly control the appropriate reverberation time to ensure speech intelligibility when using the sound public address system. According to the scale and use function of the auditorium, the mid-frequency reverberation time (fully occupied) of the auditorium should be 1.2 s±0.1 s. The measures taken include shape adjustment, through the fabric soft chair and sound absorption design a part of the wall, and the 500 Hz and 1000 Hz reverberation time estimated is 1.12 s and 1.29 s. The reverberation time (unoccupied) of the lecture hall was tested when the chair was not installed but the interior decoration was basically completed, and the average reverberation time of 500 Hz and 1000 Hz was 1.20 s and 1.30 s through the Saibin formula, and the reverberation time design of the audience hall reaches the expected index.

Keywords Lecture hall, Architectural acoustics, Reverberation time, Measured

---

环境敏感性对声景恢复效应的影响

The Effect of Environmental Sensitivity on Soundscape Restoration

高伟夫，王超，马蕙，康健
Gao Weifu, Wang Chao, Ma Hui, Kang Jian

高伟夫，天津大学建筑学院，博士研究生。
王超，天津大学建筑学院，副研究员。
马蕙，天津大学建筑学院，教授。
康健，天津大学建筑学院 / 英国伦敦大学巴特莱特建筑学部，教授。

摘要 本研究旨在探讨不同环境敏感性人群在声景心理生理恢复效应上的差异。在实验中，43 名被试通过高敏感人群量表分为高敏感性和低敏感性两组，被试在经过口算任务诱导产生一定压力后，对声景刺激进行了体验，其心理恢复反应与生理恢复反应分别通过情绪自评量表与皮肤电活动水平进行了测量。研究结果揭示了环境敏感性对于声景心理生理恢复效应的影响。这些发现可以为城市声景设计及个人声景恢复提供参考。

关键词 环境敏感性；声景；心理恢复；生理恢复；声景设计

Abstract The aim of this study was to investigate the differences in the effects of soundscapes on psychological and physiological restoration in people with different levels of environmental sensitivity. In the experiment, 43 participants were divided into two groups, high and low sensitivity, based on the Highly Sensitive Person Scale. After inducing stress through mental arithmetic tasks, the participants experienced soundscape stimuli, and their psychological and physiological restoration responses were measured using the Self-Assessment Manikin (SAM) and Electro dermal Activity (EDA), respectively. The results revealed the impact of environmental sensitivity on the effects of soundscape on psychological and physiological restoration. These findings may provide a reference for urban soundscape design and individual soundscape restoration.

Keywords Environmental sensitivity; Soundscape; Psychological restoration; Physiological restoration; Soundscape design

地毯对楼板撞击声隔声性能的影响研究

Study on the Influence of Carpet on the Sound Insulation Performance of Floor Impact

温庆韬，欧达毅
Wen Qingtao, Ou Dayi

摘要 本文以华侨大学某办公建筑的楼板为例，参考国内推荐标准与现行规范，对校内某办公建筑的楼板能进行撞击声隔声性测量与评价。在此基础上，铺设不同厚度的网格地毯与纯棉拉绒地毯，再次对楼板撞击声压级进行测量，分析铺设地毯对楼板撞击声隔声性能的影响。结果显示，该楼板的撞击声隔声性能不满足国家现行规范要求，铺设地毯能有效提升楼板撞击声隔声性能，显著降低楼板中高频撞击声压级，且 PVC 网格地毯的撞击声隔声性能优于相同厚度的纯棉拉绒地毯。

关键词 楼板撞击声；隔声性能；地毯；实测

Abstract This paper takes the floor slab of an office building of Huaqiao University as an example, and refers to the domestic recommended standards and current norms to measure and evaluate the sound insulation of the floor of an office building on campus. On this basis, PVC mesh carpets and cotton brushed carpets with different thicknesses were laid in turn, and the impact sound pressure level of the floor was analyzed, and the influence of carpet laying on the sound insulation performance of floor impact was analyzed. The results show that the impact sound insulation performance of the floor does not meet the requirements of the current national regulations, and the carpet laying can effectively improve the impact sound insulation performance of the floor slab, significantly reducing the high frequency impact sound pressure level in the floor, and the impact sound insulation performance of PVC mesh carpet is better than that of pure cotton brushed carpet of the same thickness.

Keywords Sound of floor slab impacts; Sound insulation performance; Carpet; On-site measurement

温庆韬，华侨大学建筑学院，硕士研究生。
欧达毅，华侨大学建筑学院，教授。

教室非母语言语可懂度预测模型

A Prediction Model of Non-linguistic Speech Intelligibility in Classrooms

李娜，孟琪，杨达 *，武悦
Li Na, Meng Qi, Yang Da, Wu Yue

摘要 随着我国“双一流”大学建设的推进，为提升高校国际化办学水平，非母语授课成为高校提升国际化影响力的重要举措。本文以日语为例，探究了在非母语授课背景下，教室言语可懂度随语言传输指数（STI）的变化规律，并提出日语言语可懂度预测模型。分别在 6 间教室进行了言语可懂度测试以及声学测量，共 132 名学生参与了言语可懂度测试，与此同时测量指定位置的 STI。通过逻辑回归模型分析言语可懂度得分和声学参数之间的关系。本文选择改进的 S 形回归模型作为基本回归方程，以描述 STI 对日语言语可懂度的影响规律，同时提出了基于 STI 的日语言语可懂度预测模型。

关键词 非母语授课；日语言语可懂度；语言传输指数；预测模型

Abstract With the advancement of the construction of “double first-class” universities, non-native language teaching has become an important measure for universities to enhance their international influence to improve the international level of universities. Taking Japanese as an example, this study explores the variation of speech intelligibility with speech transmission index (STI) in the context of non-native language teaching and proposes a Japanese language intelligibility prediction model. Speech intelligibility tests and acoustic measurement were conducted in 18 classrooms, and a total of 372 students participated in the speech intelligibility test, while measuring STIs at designated locations. The relationships between language intelligibility scores and acoustic parameters are analyzed by logistic regression models. In this paper, the improved “S” form regression model is selected as the basic regression equation to describe the influence of STI on Japanese language intelligibility. At the same time, a Japanese language intelligibility prediction model based on STI is proposed.

Keywords Non-native teaching; Japanese speech intelligibility; STI; Prediction model

李娜，哈尔滨工业大学人文社科学部，讲师；哈尔滨工业大学建筑学院，博士研究生。
孟琪，哈尔滨工业大学建筑学院，教授。
杨达，哈尔滨工业大学建筑学院，助理教授。
武悦，哈尔滨工业大学建筑学院，副教授。

高密度城区山体公园的声景感知

Soundscape Perception of Mountain Parks in High-density Urban Areas

张小卉，路晓东，石竹
Zhang Xiaohui,
Lu Xiaodong,
Shi Zhu

摘要 城市公园需要营造良好的声环境。高密度城区山体公园，因紧邻城市道路和自身空间特性，其声环境不同于普通城市公园。本研究以大连富国公园为例，利用 VR 技术和声音回放技术，通过声景评价进行比较，分析声源感知程度、声景感知维度和声景恢复性。结果表明，受城市交通噪声影响，高密度城区山体公园不同山体部位的声景感知具有显著差异；基于 4 种声源感知对于愉悦感、事件感和声景恢复性的预测模型发现，交通噪声感知对愉悦感、事件感和声景恢复性的影响均最大。研究结果将对高密度城区山体公园的游览规划设计起到趋利避害的作用。

关键词 山体公园；高密度城区；声景感知；声景恢复性；交通噪声

Abstract Urban parks need a good acoustic environment. The acoustic environment of high-density urban mountain parks is different from ordinary ones due to their proximity to urban roads and own spatial characteristics. Taking Dalian Fuguo Park as an example, the study uses VR and sound playback technology to compare soundscape evaluation, including sound source perception, soundscape perception dimension and soundscape restoration. The results show that, affected by urban traffic noise, there are significant differences in the soundscape perception of different mountain parts in mountain parks; Based on the prediction models of four sound source perceptions for pleasant, eventful and soundscape restoration, it is found that traffic noise perception has the greatest impact on pleasant, eventful and soundscape restoration. The research results will play a role in seeking advantages and avoiding disadvantages for the tourism planning and design of mountain parks in high-density urban areas.

Keywords Mountain park; High-density urban; Soundscape perception; Soundscape restoration; Traffic noise

张小卉，大连理工大学，硕士研究生。
路晓东，大连理工大学，博士，教授。
石竹，大连理工大学，硕士研究生。

英国 Robust Details 建筑隔声性能评价体系介绍与思考

Introduction and Consideration of the Evaluation System of Sound Insulation Performance: Robust Details in British

赵启元，闫国军，刘青云
Zhao Qiyuan, Yan Guojun, Liu Qingyun

摘要 介绍英国 Robust Details 建筑隔声性能评价系统，该系统针对不同产品、构造在现场实施过程中的细节把控，通过多次不同项目的测试结果比对，提高产品、构造在不同施工环境下的稳健性，从而提高产品、构造在不同项目中的声学可靠性。本文涵盖该系统发展历程、申请方式、评价方法、控制手段，达成的效果等方面，并结合我国现状，提出了一些想法和思考。

关键词 Robust Details；隔声性能；现场检测；声学可靠性

Abstract This paper introduces the British Robust Details - sound insulation performance evaluation system, which aims to control the details of different products and structures in the field implementation process, and improve the soundness of products and structures in different construction environments by comparing the test results of many different projects, so as to improve the acoustic reliability of products and structures in different projects. This paper covers the development history, application method, evaluation method, control means, and achieved results of the system. In combination with the current situation in China, some ideas and thoughts are put forward.

Keywords Robust Detail; Sound insulation; Field test; Acoustic reliability

赵启元，中国建筑科学研究院有限公司，工程师。
闫国军，中国建筑科学研究院有限公司，正高级工程师，声学室主任。
刘青云，中国建筑科学研究院有限公司，助理工程师。

新噪声法下的住宅建筑隔声降噪措施探讨

Discussion on Sound Insulation and Noise Reduction Measures of Residential Buildings Under the New Noise Law on Prevention and Control of Noise Pollution

林琳，王沁芳，陈瑞，彭芃，张鹏东

Lin Lin, Wang Qinfang, Chen Rui, Peng Peng, Zhang Pengdong

林琳，陈瑞，彭芃，常州市建筑科学研究院集团股份有限公司，助理工程师。

王沁芳，常州市建筑科学研究院集团股份有限公司，研究员级高级工程师。

张鹏东，江苏尼高科技有限公司，助理工程师。

摘要 自 2022 年 6 月 5 日起实施的《中华人民共和国噪声污染防治法》对噪声污染治理提出了更高的管理与噪声防治要求。本文根据实验室中不同型号建筑外窗隔声性能检测数据，分析了目前常用建筑外窗的性能差异，进而提出室内声环境优化建议。结合实验室的隔声试验结果，基于《建筑环境通用规范》（GB 55016—2021）要求提出了有关住宅建筑隔声降噪措施建议。

关键词 噪声法；住宅建筑声环境；建筑外窗隔声性能；降噪措施

Abstract *The Law of the People's Republic of China on the Prevention and Control of Noise Pollution*, which came into effect on June 5, 2022, puts forward higher management and noise prevention requirements for noise pollution control. Based on the test data of sound insulation performance of exterior windows of different types of buildings in the laboratory, this paper analyzes the performance differences of exterior Windows of commonly used buildings at present, and then puts forward suggestions for optimizing the indoor sound environment. Based on the requirements of *General Code for Building Environment* (GB 55016—2021), suggestions on sound insulation and noise reduction measures for residential buildings are put forward.

Keywords Law on prevention and control of noise pollution; Residential building sound environment; Sound insulation of window; Noise reduction measure

---

老年群体对城市开放空间声景评价研究概述

An Overview of the Research on Soundscape Evaluation of Urban Open Space by the Elderly

谭若晨，欧达毅 *

Tan Ruochen, Ou Dayi

谭若晨，华侨大学建筑学院，硕士研究生。

欧达毅，华侨大学建筑学院，教授，博士生导师。

摘要 目前的声景评价研究中，针对老年群体的声景评价研究相对较少。本文在参考国内外文献的基础上，对现有的文献进行系统综述，分析老年人较其他群体在声舒适度、声元素喜好度及主观响度评价方面存在的差异。结果表明：城市开放空间中，在声舒适度方面，多数研究发现不同年龄之间存在显著差异，其中老年人往往是最满意的群体；在声元素喜好度方面（已得到较为一致性的结论），老年人相对年轻人更喜欢自然声，且对鸟叫声的喜好程度最高；在主观响度评价方面，前人研究尚未形成统一的评价结果。

关键词 老年人；城市开放空间；声景；评价指标

Abstract At present, in the studies on the soundscape evaluation of different groups, there are relatively few studies on the soundscape evaluation of the elderly groups. Based on the reference of domestic and foreign literature, this paper systematically reviews the existing literature, and analyzes the differences between the elderly and other groups in the evaluation of acoustic comfort, acoustic element preference and primary loudness. The results show that: in terms of acoustic comfort in urban open space, there is a significant difference between age and acoustic comfort in most studies, and the elderly are usually the most satisfied group. In terms of the preference for sound elements (a relatively consistent conclusion has been reached), old people prefer natural sounds more than young people, and the preference for bird calls is the highest. In the aspect of subjective loudness evaluation, previous studies have not yet formed unified evaluation results.

Keywords The elderly; Urban open space; Soundscape; Evaluation index

工业遗产创意园声环境研究——以广州 T.I.T 创意园为例

Study on Acoustic Environment of Industrial Heritage Creative Park —Taking T.I.T Creative Park in Guangzhou as an Example

何小钰，王红卫*，周昕彤
He Xiaoyu, Wang Hongwei, Zhou Xintong

摘要 工业遗产创意园作为工业遗产保护与再利用的主要形式，其声环境品质对于创意园整体环境质量的提升有着重要的作用。目前国内关于工业遗产创意园声环境的研究极少。本文以广州 T.I.T 创意园为研究对象，对园区的声环境展开实地测量和主观问卷调查。研究结果表明，园区内几处主要休闲广场声环境明显超出了我国现行声环境质量标准限值；在主观评价上，创意园整体声环境满意度和大众重视程度都较高，其中自然声的声好感度和环境协调度评价均值最高。最后就园区声环境现状提出一些优化策略。

关键词 工业遗产创意园；声环境；主观问卷；满意度

Abstract As the main form of industrial heritage protection and reuse, the acoustic environment quality of industrial heritage creative parks plays an important role in improving the overall environmental quality of creative parks. At present, there is few research on the acoustic environment of industrial heritage creative parks in China. Taking Guangzhou T.I.T Creative Park as the research object, this paper conducted field measurement and subjective questionnaire survey on the acoustic environment of the park. The results showed that the acoustic environment of several main leisure plazas in the park obviously exceeds the limits of the existing standards of acoustic environment quality. In terms of subjective evaluation, the overall sound environment satisfaction and public attention were high, among which the sound preference for natural sound and environmental coordination degree were the highest. Finally, the paper put forward some optimization strategies for the acoustic environment of the park.

Keywords Industrial heritage creative park; Acoustic environment; Subjective questionnaire; Satisfaction degree

何小钰，华南理工大学建筑学院，硕士研究生。
王红卫，华南理工大学建筑学院 / 亚热带建筑科学国家重点实验室，教授。
周昕彤，华南理工大学建筑学院，硕士研究生。

南音在城市广场中的心理恢复性效应研究

A Study on the Psychological Restorative Effect of Nanyin in Urban Squares

陈岩君，欧达毅*
Chen Yanjun, Ou Dayi

摘要 南音是我国现存最古老的乐种之一，于 2009 年入选世界非物质文化遗产名录。本文以南音纯乐器曲为例，在实验室中借助短版修订恢复量表（SRRS）研究纯南音环境、南音 + 广场结合的环境下的心理恢复效果。本实验招募 19 名被试，随机分成 2 组分别在纯南音环境和视听环境下进行实验，相关结论可以为南音的传承与保护工作和声景恢复性研究提供参考。

关键词 南音；声景；恢复性；城市广场

Abstract Nanyin is one of the oldest surviving music genres in China and was inscribed on the World Intangible Cultural Heritage List in 2009. In this paper, we use the short version of the Revised Recovery Scale (SRRS) in the laboratory to study the psychological recovery effects in the pure Nanyin environment and the combined Nanyin and square environment. Nineteen subjects were recruited and randomly divided into 2 groups to conduct the experiment in a pure Nanyin environment and an audiovisual environment respectively. The relevant findings can provide a reference for the inheritance and conservation work of Nanyin and soundscape restorative research.

Keywords Nanyin; Soundscape; Restorative; Urban square

陈岩君，华侨大学建筑学院，硕士研究生。
欧达毅，华侨大学建筑学院，教授，博士生导师。

深圳玛丝菲尔秀场声学设计策略研究

Research on Acoustic Design Strategy of Shenzhen Masefield Show

周昕彤，王红卫，何小钰
Zhou Xintong, Wang Hongwei, He Xiaoyu

周昕彤，华南理工大学，硕士研究生。
王红卫，华南理工大学，教授。
何小钰，华南理工大学，硕士研究生。

摘要 秀场作为独特的音质空间，兼顾多种功能要求。本文以深圳玛丝菲尔秀场为例，通过 Odeon 软件模拟分析秀场声场特点，研究秀场类音质设计策略。结果表明秀场声场复杂，混响时间偏长，存在大量声缺陷。根据室内装饰效果要求选择声学材料，改善秀场空间的音质。研究结果可为秀场类空间的音质设计和研究提供参考。

关键词 秀场；声学设计策略；声场特点；建筑声学

Abstract As a unique sound quality space, the show takes into account a variety of functional requirements. This paper takes the Shenzhen Masfil Show as an example, analyzes the characteristics of the show sound field with Odeon software simulation, and studies the design strategy of show sound quality. The results show that the show sound field is complex, the reverberation time is long, and there are a large number of sound defects. Acoustic materials are selected according to the requirements of interior decoration effects to improve the sound quality of the show space. The research results can provide a reference for the sound quality design and research of show space.

Keywords Show; Acoustic design strategy; Sound field characteristics; Architectural acoustics

---

城市社区公园声景感知研究

Research on Soundscape Perception in Urban Community Parks

韩振坤，唐征征，吕海景，崔小元
Han Zhenkun, Tang Zhengzheng, L ü Haijing, Cui Xiaoyuan

韩振坤，哈尔滨工业大学建筑学院，副教授。
唐征征，哈尔滨工业大学建筑学院，讲师。
吕海景，东北农业大学艺术学院，副教授，环境设计系主任。
崔小元，东北农业大学艺术学院环境设计系，硕士研究生。

摘要 社区公园作为贴近日常生活的景观体验空间对市民生活具有非常重要的作用。本文以声景感知为视角，旨在了解人们对城市社区公园声景的感知特征，以为相关设计与管理提供参考。本研究通过对使用者的声景感知及其影响因素，使用者的声喜好差异性进行分析，结果表明，大多数使用者最先感知到声喜好较低的道路交通声，而对声喜好较高的自然声的感知率较低；年龄与行为方式均使声喜好发生规律性变化。

关键词 社区公园；声景；声喜好；感知

Abstract As a landscape experience space close to daily life, community park plays a very important role in citizens' life. From the perspective of soundscape, this paper aims to understand people's perception of soundscape in urban community parks, and provide references for related design and management. This study analyzed users' perception of soundscape and its influencing factors, as well as differences in users' acoustic preferences. The results showed that most users first perceived the road traffic noise they did not like, while the perception rate of natural sound with high acoustic preference was lower. Age and behavior patterns cause regular changes in vocal preference.

Keywords Community park; Sound scene; Acoustic preferences; Perception

基于建筑信息模型的绿色建筑隔声评价技术与应用

Evaluation of Green Building Sound Insulation Based on Building Information Models

杨玉锦，殷明刚，杨玥，
刘奕彪，赵敬源 *
Yang Yujin, Yin Minggang, Yang Yue, Liu Yibiao, Zhao Jingyuan

杨玉锦，长安大学建筑学院，博士研究生，工程师。
殷明刚，上海华东建筑设计研究院有限公司，建筑性能分析工程师。
杨玥，江苏省建筑科学研究院有限公司，智慧城市研究中心副主任，高级工程师。
刘奕彪，江苏省建筑科学研究院有限公司，工程师。
赵敬源，长安大学，教授，博士生导师。

摘要 绿色建筑设计标准对建筑隔声的要求不断提升，对不同类型建筑空间的隔声性能评价规则也同步细化，传统的手工计算验证方法工作量大、效率低下且容易发生误操作，在实际工作中难以应对新的需求。基于建筑信息模型，利用经过严格验证的物理模型开展建筑隔声计算有助于提高计算准确性，并提高效率。本文基于建筑信息模型技术建立了一套符合开放理念和国际标准的绿色建筑隔声评价的基本工作框架，在此框架下依托 AcouBAT 工具实现了对某绿色建筑示范项目的多功能、多区域房间的建筑隔声评价。

关键词 建筑隔声；建筑信息模型；性能分析；协同设计

Abstract The requirements of green buildings for building sound insulation have been continuously improved, and the evaluation rules for the sound insulation performance of different types of building spaces have also been refined simultaneously. The traditional manual checking method has a large workload, and it is difficult to cope with new requirements in actual work. Based on the building information model, it is helpful to improve the accuracy and efficiency of the building sound insulation calculation by using a strictly verified physical model and empirical formula. Based on the building information modeling technology, this paper establishes a basic framework for the sound insulation evaluation of green buildings that conforms to the open concept and international standards. In this framework, based on the AcouBAT tool, the building sound insulation evaluation of multi-function and multi-area rooms of a green building demonstration project is realized.

Keywords Building sound insulation; Building information model; Performance analysis; Collaborative design

---

长江沿岸城市公共开放空间声环境营造

Sound Environment Creation in Public Open Spaces of the Yangtze River City

范波，李运江 *，许华华
Fan bo，Li Yunjiang，Xu Huahua

范波，三峡大学建筑技术科学研究所，实验师。
李运江，三峡大学建筑技术科学研究所，教授。
许华华，三峡大学建筑技术科学研究所，讲师。

摘要 为提高武汉长江城市阳台的辨识度，感受武汉特有的地理、历史和文化特征，提出了 CIS 声环境设计理念，即听闻舒适性（Comfort）、辨识度（Identification）和标志性（Symbolics），并以此为目标，采用问卷调查、计算机仿真和物理实验等方法，对场地声景元素进行梳理，并总结出普遍适用于长江沿岸室外开放空间的声环境设计方法。

关键词 武汉月亮湾城市阳台；声环境营造；声景舒适性；辨识度和标志性

Abstract In order to improve the identification of the Wuhan Yangtze River City Balcony, We proposed the CIS soundenvironment design concept, i.e., Auditory comfort, Identification and Symbolics. We use questionnaire surveys, computer simulations and physical experiments to sort out soundscape elements, and to summarise sound environment design methods that are generally applicable to outdoor open spaces along the Yangtze River.

Keywords Wuhan Moon Bay City Balcony; Sound environment creation; Sound environment comfort; Identification and Symbolics

古戏台中的哈斯效应实践

——兼议鼓板指挥何以成为可能

The Practice of the Haas Effect in the Ancient Theatre—Another discussion on what makes drumboard conducting possible

杨阳，翟晓多，高策，任世君
Yang Yang, Zhai Xiaoduo, Gao Ce, Ren Shijun

摘要 本文发现明正德年间创建的新绛县稷益庙戏台的通面阔超过 17.7 m，其后建造的明代戏台最大尺寸也仅为其 82%，即便是在场上演员数量达到鼎盛时期的清代也很少有超越它的。主观评价结果表明，在该戏台上演出时有“拉不动”和“对不齐”的情况，声学测量与分析结果证明，通面阔太大是造成上述不良主观感受的“元凶”，重要的是，明正德年间，哲匠们已通过营造这座戏台认识了哈斯效应，并将其经验教训成功应用在其后营造的其他戏台上，科学消除了延迟和回声对戏曲表演的干扰，解决了鼓板兼职戏曲乐队指挥时文武场声部协调统一的问题。

关键词 明代戏台；延迟；回声；声学技术；哈斯效应

Abstract This paper finds that the through-face width of the Jiyi Temple Theatre in Xinjiang County, which was built during the Zhengde period of the Ming Dynasty, exceeds 17.7 m. The maximum size of the theatres built in the Ming Dynasty after that was only 82% of that size, and few of them exceeded that size even in the Qing Dynasty, when the number of performers in the theatre was at its peak. Subjective evaluation results show that in the performance of the theatre on the “pull not” and “alignment” of the situation, acoustic measurements and analysis results proved that the through the surface is too wide to cause the above bad subjective feelings of the “culprit! ”, importantly, during Ming Zhengde reign, the philosophers have been through the creation of this theatre to understand the Haas effect, and its lessons successfully applied in the subsequent creation of other theatre, the scientific elimination of the delay and echo on the interference of the performance of the opera, solved the drums and boards part-time theatre orchestra conductor of the sound coordination and unity of the civil and military part of the problem, which can be described as the earliest experiments of the Haas effect in China.

Keywords Ming dynasty theatre; Delay; Echo; Acoustic techniques; Haas effect

杨阳，山西大学科学技术史研究所，教授，博士生导师。
翟晓多，山西大学音乐学院，博士研究生。
高策，山西大学科学技术史研究所，教授，博士生导师。
任世君，山西大学科学技术史研究所，博士研究生。

声景对儿童参与社交互动的影响

Effects of Soundscape on Children's Participation in Social Interactions

苏婳，马蕙 *
Su Hua, Ma Hui

摘要 声景研究在提高城市空间质量方面已经取得了一定成果，但几乎没有研究关注声景对个体发展的促进作用。因此，本研究调查了声景对儿童参与社交互动的影响，为物理环境支持儿童社会化提供证据。研究在实验室中进行，通过虚拟演示的方式收集儿童参与社交互动相关数据。实验声为八种城市典型声源和控制条件（无声音），分别结合学校操场或城市公园的视觉场景呈现。结果表明，儿童声和成人声能够显著提升儿童参与社交互动的程度。尽管音乐声、动物声、和自然声能够显著提高儿童对学校操场和城市公园的使用率，但是对儿童参与社交互动没有显著的促进作用。

关键词 声景；学龄儿童；社交互动；虚拟演示

Abstract Soundscape research has achieved some results in improving the quality of urban spaces, but few studies have examined how soundscapes encourage individual development. Thus, this study investigated how soundscapes influence children' s social interactions to provide evidence that physical environments are supportive of socialization. The study was conducted in a laboratory, and children's participation in social interactions was measured through virtual demonstrations. The experimental stimuli were eight typical urban sound sources and a control condition (no sound), which were presented in combination with visual recordings of a schoolyard or urban park. The results showed that children's sounds and adult' s sounds could significantly promote children's participation in social interactions. Although classical music, animal sounds, and nature sounds could significantly attract children to use schoolyards and urban parks, they had no significant effect on promoting children's social interactions.

Keywords Soundscape; School-age children; Social interaction; Virtual demonstrations

苏婳，天津大学，博士研究生。
马蕙，天津大学，教授，博士生导师。

基于树干吸声系数的城市街旁绿地树种筛选研究

Tree Species Selection for Urban Roadside Green Spaces Based on Sound Absorption Coefficient of Trunks

李朦朦，孟琪 *
Li Mengmeng, Meng Qi

摘要 街旁绿地对改善交通噪声污染具有重要作用，然而现存研究忽略了树种间差异在不同频率噪声控制中的重要作用，难以满足城市交通噪声有效防控的需求。基于此，本研究以植物树干为切入点，针对交通噪声频谱特征系统测量 13 种树种的吸收作用，并模拟降噪效果。研究首先测得树干吸声系数的频谱特征，确定了树干主要吸声频率范围与交通噪声频率分布具备较好的一致性；其次从树种视角分析影响树干吸声效果的主要因素；最后通过模拟对比分析 15 米宽绿地种植不同树种的降噪效果。

关键词 噪声控制；交通噪声；吸声系数；自然降噪途径

Abstract Roadside green spaces play an important role in improving traffic noise pollution. However, existing studies ignore the important role of differences between tree species in the process of noise control at different frequencies, making effective urban traffic noise prevention and control difficult to meet. Therefore, this study begins with plant trunks, proposes a method for quantifying trunk and bark roughness, measures the absorption of 13 tree species for the traffic noise spectrum characteristics system, and simulates the noise reduction effect. First, the spectral characteristics of the trunk sound absorption coefficient were measured, and the trunk' s main absorption frequency range was determined to be in good agreement with the frequency distribution of traffic noise. Second, this study examines the primary factors that influence the sound absorption effect of tree trunks from the perspective of tree species. Finally, analyze the noise reduction effects of planting different tree species in a 15-meter-deep green space using simulation.

Keywords Noise control; Traffic noise; Sound absorption coefficient; Natural means for noise abatement

李朦朦，哈尔滨工业大学建筑学院，助理研究员。
孟琪，哈尔滨工业大学建筑学院，教授，博士生导师。

MLS 可变吸声扩散体应用于高尔夫 KTV 会所的初探

A Preliminary Research on the Acoustic Design of MLS Variable Acoustic Diffuser Applied to the Golf KTV Club

王波，李昱庭，戴文林，李立新，林跃华，杨思瀚，赵汉清
Wang Bo, Li Yuting, Dai Wenling, Li Lixin, Ling Yuehua, Yang Sihan, Zhao Hanqing

摘要 本文针对高尔夫 KTV 会所平行侧墙，采用发明专利 MLS 可变吸声扩散体与 MLS 吸声扩散体相互搭配，形成可变吸声扩散界面，并结合吸声窗帘等相关技术措施，提高声场均匀度，努力营造出可变混响声学空间，以便拓展 MLS 可变吸声扩散体的研究应用，为完善高尔夫 KTV 会所的声学设计提供借鉴。

关键词 高尔夫 KTV 会所；MLS 可变吸声扩散体；可变混响

Abstract In this paper, the paralled side walls of the Golf KTV Club are used and the patented MLS variable sound-absording diffuser is used to match with the sound-absorbing diffuser to form a variable sound absorbing diffusion interface, and combined with sound-absorbing curtains and other related technical measures to improve the uniformity of the sound field, strive to create a variable reverberation acoustics space. In order to expand the research and application of MLS variable sound-absorbing diffuser, and provide a reference for improving the design of the Golf KTV Club.

Keywords Golf KTV Club; MLS variable sound-absorbing diffuser; Variable reverberation

王波，厦门大学建筑与土木工程学院，博士，副教授，硕士生导师。
李昱庭，厦门大学建筑与土木工程学院，硕士研究生。
戴文林，厦门永竹竹业科技有限公司，总经理。
李立新，厦门大学建筑与土木工程学院，教授，厦门大学建筑设计研究院，院长。
林跃华，厦门大学建筑与土木工程学院，工程师。
杨思翰，厦门大学建筑与土木工程学院，本科生。
赵汉清，厦门大学建筑与土木工程学院，硕士研究生。

寒地冬季高校室内公共空间声喜好初探

Preliminary Study on the Sound Preference of Indoor Public Spaces in Winter Universities in Cold Regions

陈昱，唐征征，张卷舒 *
Chen Yu,
Tang Zhengzheng,
Zhang Juanshu

摘要 舒适的校园声环境十分有利于高校人群的学习、工作，声喜好是评价声环境的一个重要方面。本文采用问卷调查法和语义细分法，选取哈尔滨工业大学建筑学院教学楼作为研究对象，对寒地冬季高校人群的声喜好进行研究。结果表明：由于寒地建筑特点，冬季室外大部分自然声被隔绝，室外不喜好声音的不良影响大幅降低。在室内公共空间中，声源种类以人为声为主，交谈声占据重要地位，适当的乐声是较受人群喜爱的，人们对运动噪声、设备运行声感到厌恶。基于此，本文进一步提出相应的改善措施。

关键词 寒地；冬季；高校；室内公共空间；声喜好

Abstract A comfortable campus sound environment is very conducive to the learning and work of the university population, and voice preferences are an important aspect of evaluating the sound environment. In this paper, questionnaires and semantic subdivision methods are used to select the teaching building of the School of Architecture of Harbin Institute of Technology as the research object to study the sound preferences of college students in cold winter. The results indicate that due to the characteristics of cold buildings, most of the natural sound outside is isolated in winter, and the adverse effects of outdoor dislike of sound are significantly reduced. In indoor public spaces, human noise is the main type of sound source, and conversation sound plays an important role. Appropriate music is more popular among people, and people feel disgusted with sports noise and equipment operation sound. Based on the results, the article further proposes corresponding improvement measures.

Keywords Cold regions; Winter; Universities; Indoor public spaces; Voice preference

陈昱，哈尔滨工业大学建筑学院，本科生。
唐征征，哈尔滨工业大学建筑学院，讲师。
张卷舒，哈尔滨工业大学建筑学院，副教授。

街道高宽比对临街立面交通噪声的影响预测

Prediction of the Influence of Street Aspect Ratio on the Acoustic Environment of Street-Facing Facades

邹杨，路晓东
Zou Yang, Lu Xiaodong

摘要 街道几何形态对城市声环境存在不可忽视的影响。为了研究高宽比对临街立面交通噪声的影响，以声压级 (SPL)、混响时间 (T30) 和早期衰减时间 (EDT) 为评价指标，利用 Odeon 软件进行声学仿真模拟，在建筑高度固定为 30 米的情况下，进行了在不同街道宽度（H/W 分别为 0.75、1、1.5、3）影响下的交通噪声的空间分布比较。结果表明：立面声压级与 H/W 呈负相关，随着高宽比的减小，声压级曲线转折点增大；当高宽比增大时，立面反射声能量占比增大； EDT 和 T(30) 与高宽比呈负相关。

关键词 街道高宽比；立面声环境；声学仿真；建筑高度；街道宽度

Abstract The geometry of the street has a non-negligible influence on the urban sound environment. In order to study the influence of aspect ratio on the traffic noise of the street façade, the acoustic simulation index (SPL), reverberation time (T30) and early attenuation time (EDT) were taken as the evaluation index, the acoustic simulation was carried out by Odeon software, the building height was fixed at 30 in the case of meters, and the spatial distribution of traffic noise under the influence of different street widths (H/W was 0.75, 1, 1.5, 3, respectively) was compared. The results show that the sound pressure level of the façade is inversely proportional to H/W, and the turning point of the sound pressure level curve increases with the decrease of the aspect ratio. When the aspect ratio increases, the proportion of sound energy reflected by the façade increases; EDT and T(30) are inversely proportional to the aspect ratio.

Keywords Street aspect ratio; Façade acoustic environment; Acoustic simulation; Building height; Street width

邹杨，大连理工大学建筑与艺术学院，硕士研究生。
路晓东，大连理工大学建筑与艺术学院，教授，博士生导师。

住宅室内健康声环境的要素和概念框架

Elements and a Conceptual Framework of Healthy Indoor Acoustic Environments in Residential Buildings

陈凯，康健，马蕙 *
Chen Kai, Kang Jian, Ma Hui

陈凯，天津大学建筑学院，博士研究生。
康健，英国伦敦大学学院巴特莱特建筑学部，教授；天津大学建筑学院，教授，博士生导师。
马蕙，天津大学建筑学院，教授，博士生导师。

摘要 本研究旨在调查居民对住宅室内健康声环境的评价、理解和需求，基于扎根理论的研究方法，对 35 名受访者的访谈记录进行了编码。研究结果提出了住宅室内健康声环境的要素以及一个用于描述要素之间关系的概念框架。在所有要素中，依恋性、私密性、自主性和可控性被居民认为是住宅室内健康声环境的特征（或需求）。本研究为健康住宅中的声环境设计提供了理论参考。

关键词 健康声环境；室内声环境；健康住宅；扎根理论；概念框架

Abstract The study aims to investigate residents' evaluation, understanding and requirements for healthy indoor acoustic environments in residential buildings. The grounded theory was used to code the interview transcripts of 35 respondents. The findings indicated the elements of healthy indoor acoustic environments in residential buildings and a conceptual framework that described the relationships between the elements. Among all elements, Attachment, Privacy, Autonomy, and Controllability were identified as four characteristics (or requirements) of healthy indoor acoustic environments in residential buildings. The study provided a theoretical reference for the design of acoustic environments in healthy housing.

Keywords Healthy acoustic environment; Indoor acoustic environment; Healthy housing; Grounded theory; Conceptual framework

---

面向线上教学的智慧教室声光环境设计

Design of Smart Classroom Sound and Light Environment for Online Teaching

徐炤圆，张三明，楼志坚
Xu Zhaoyuan, Zhang Sanming, Lou Zhijian

徐炤圆，浙江大学建筑工程学院，硕士研究生。
张三明，浙江大学建筑工程学院，副教授。
楼志坚，浙江大学建筑工程学院，硕士研究生。

摘要 随着新兴科技与教学技术的发展，教室出现了直播录播、线上教学、远程互动等多种新需求，智慧教室应运而生。高质量的课程画面与音质效果需求，对智慧教室声环境与光环境提出了更高的要求。本文梳理了智慧教室的概念与特点，针对智慧教室的新需求，借鉴了演播室与录音室等的设计思路，提出了智慧教室声光环境设计的目标，并从多个方面给出了智慧教室声光环境设计优化的思路与具体措施，为智慧教室的建设实践提供指导。

关键词 智慧教室；远程互动教学；直播录播；教室声环境；教室光环境

Abstract With the development of emerging technology and teaching technology, classrooms have emerged with various new demands such as live streaming, online teaching, and remote interaction, and then smart classrooms. The demand for high-quality course visuals and sound effects places higher demands on the sound and light environments of smart classrooms. This article summarizes the concept and characteristics of smart classrooms, and in response to the new needs of smart classrooms, draws on design ideas such as studios and recording rooms, and proposes the goal of designing the sound and light environment of smart classrooms. And provided ideas and specific measures for optimizing the sound and light environment design of smart classrooms from multiple aspects, providing guidance for the construction practice of smart classrooms.

Keywords Smart classroom; Distance interactive teaching; Live and recorded broadcasting; Acoustic environment of classroom; Lighting environment of classroom

第 3 篇

建筑采光与照明

节律健康视角下教室天然光和人工光组合模式研究

Study on Daylight and Electric Light Combination Modes in Classrooms from the Perspective of Circadian Rhythm

童蝶宏，徐俊丽*，郝慧羽
Tong Diehong, Xu Junli, Hao Huiyu

摘要 在“健康中国”背景下，改善教室光环境对促进学生的健康节律至关重要。然而在室内环境中，天然光所提供的光照量难以满足节律健康需求。本研究通过 DIALux evo 软件构建苏州地区典型教室光环境仿真模型，当仅天然光、仅人工光时，分析教室中不同区域节律效应量化指标 melanopic EDI 的变化规律。基于节律健康需求，提出优化天然光和人工光组合模式的方法，未来将进一步建立可预测空间与时间 melanopic EDI 的简化模型，为教室健康光环境设计提供参考。

关键词 节律健康；教室；视黑素等效日光（D65）照度；天然光；人工光；组合模式

Abstract In the context of “Healthy China”, improving the classroom light environment is essential to promote students' circadian rhythm. However, in indoor environments, the amount of light provided by daylight is not sufficient to meet the needs of the circadian rhythm. In this study, a simulation model of a typical classroom light environment in Suzhou was constructed by DIALux evo to analyze the change rules in melanopic EDI, a quantitative index of rhythm effects in different areas of the classroom, in the circumstances of only daylight and only electric light. Based on the needs of circadian rhythm, a method to optimize the combination modes of daylight and electric light was proposed. Furthermore, in the future, a simplified model to predict spatial and temporal melanopic EDI was estimated to provide a reference for classroom healthy light environment design.

Keywords Circadian rhythm; Classroom; Melanopic equivalent daylight illuminance; Daylight; Electric light; Combination modes

童蝶宏，苏州大学建筑学院，硕士研究生。
徐俊丽，苏州大学建筑学院，副教授，硕士生导师。
郝慧羽，苏州大学建筑学院，硕士研究生。

利用“眼动－视野”模型的室内人眼光暴露测算方法

Calculation and Measurement of Light Exposure of Human Eyes in Indoor Space Based on the Gaze-dependent Visual Field Model

何思琪，严永红
He Siqi, Yan Yonghong

摘要 研究提出将全视野瞳孔进光量（以下称“入眼光通量”）作为人眼光暴露水平的量化指标，综合考虑人眼的注视行为、全视野范围和瞳孔光反应三者的影响，建立“眼动－视野”模型并推导瞳孔表观面积的计算方法，并开展验证实验实施一组算例。结果显示，入眼光通量的空间接收效率与瞳孔椭圆率、光线入射角有关。研究完成了入眼光通量的初步测算流程，为人本照明系统和光环境质量评估提供了技术支持。

关键词 人眼光暴露；入眼光通量；注视角；瞳孔表观面积

Abstract This study proposed to use the luminous flux entering the human eye from the whole visual field as a metric to quantify the light exposure level of human eyes, considering human gaze behavior, visual field, and pupillary light reaction. Meanwhile, a gaze-dependent visual field model was established, and the calculation of pupillary apparent area was induced. Then, a verification test was implemented assuming a Ganzfeld luminous environment. As a result, the spatial receiving efficiency of luminous flux entering human eyes is correlated with the pupil diameter ratio and incident angle of incoming light. In conclusion, this study proposed and validate the calculation and measurement of luminous flux entering human eyes based on the gaze-dependent visual field model and would help to improve the human-centric lighting system and the quality evaluation of luminous environment.

Keywords Light exposure of human eyes; Luminous flux entering eyes; Gaze angle; Pupillary apparent area

何思琪，南京林业大学风景园林学院，讲师。
严永红，重庆大学建筑城规学院，教授。

使用 BSDF 数据的教室黑板面亮度对比度的模拟分析

Assessing the Luminance Contrast Condition within the Blackboard of a Classroom Using BSDF

遇大兴，陈亚楠，边宇 *
Yu Daxing, Chen Yanan, Bian Yu

摘要 黑板面上出现的反射光是导致教室中视觉不适的主因，本文的目的是探讨黑板内亮度对比度在教室采光设计评价中的适用性，包括：在模拟时使用 BSDF（双向散射分布函数）数据并提出基于对比度的评价模型，验证模型并对提议的 CFS（复杂采光系统）进行评估。结果显示：目前设置的对比度模拟的误差在 17% 以内；在教室中使用 DGP 将低估眩光条件；经设计后的百叶可以在显著改善对比度状况的同时保留日光的可用性。本研究结论为：① BSDF 数据用于模拟教室采光系统准确度高于传统采光模拟方法；②黑板内亮度对比度在教室采光设计评价中具有适用性。

关键词 视觉舒适度；亮度对比度；BSDF；评价模型；采光设计；教室

Abstract Reflected light appearing on the blackboard surface is the main cause of visual discomfort in the classroom. The purpose of this paper is to investigate the applicability of the luminance contrast within the blackboard in the evaluation of classroom lighting design, including using BSDF data in the simulation and proposing a contrast-based evaluation model, validating the model and evaluating the proposed CFS (complex lighting system). The results show that: the error of the currently set contrast simulation is within 17%; the use of DGP in classrooms will underestimate glare conditions; and the designed louvers can retain daylight availability while significantly improving contrast conditions. This study concludes：① the accuracy of daylight simulation of the fenestration system in the classroom using BSDF data is higher than the traditional method; ② the contrast of luminance within the blackboard is applicable in the evaluation of classroom daylighting design.

Keywords Visual comfort; Luminous contrast; BSDF; Evaluation model; Daylighting design; Classroom

遇大兴，华南理工大学建筑学院，副教授。
陈亚楠，华南理工大学建筑学院，硕士研究生。
边宇，华南理工大学建筑学院，副教授。

基于运动行为的高校体育馆眩光点位分析研究

Research on Glare Point Analysis of University Gymnasium Based on Sports Behavior

范晴，张明宇 *，刘彦汝
Fan Qing, Zhang Mingyu, Liu Yanru

摘要 由于高校体育馆内使用者运动位置角度的不确定性，导致天然采光不舒适眩光研究复杂。调研发现高校体育馆打羽毛球、篮球使用频率最高，基于这两种运动行为高频活动点位及视看角度，通过“观察—还原—统计”研究方法得出高校体育馆典型眩光点位 28 个（其中羽毛球场地 12 个，篮球场地 16 个）和视看角度，补充均匀布点法无法适应体育馆内不同类型场地活动的空缺，以及传统眩光模拟常选用垂直角度为 0° 进行眩光模拟在动态活动中的不足。

关键词 高校体育馆；不舒适眩光；运动行为；眩光模拟

Abstract Due to the uncertainty of the user's movement position and angle in the university gymnasium, the research on natural lighting uncomfortable glare is complicated. The survey found that users play badminton and basketball in university gymnasiums most frequently. Based on the high-frequency activity points and visual angle of these two sports behaviors, 28 typical glare points (including 12 badminton courts and 16 basketball courts) and visual angle of university gymnasiums were obtained through the research method of "observation reduction statistics", to supplement the gap that the uniform distribution method cannot adapt to different types of field activities in the gymnasium, And the traditional glare simulation often selects a vertical angle of 0° for glare simulation in dynamic activities.

Keywords University gymnasium; Uncomfortable glare; Sports behavior; Glare Simulation

范晴，天津大学，博士研究生。
张明宇，天津大学，副教授，建筑技术科学研究所副所长。
刘彦汝，天津大学，硕士研究生。

内蒙古地区城镇夜空广域光现状分析报告

Analysis of the Current Situation of Town Night Sky Wide-area Light in Inner Mongolia

丁杨，曹敦，魏炜，韩则阳，苏晓明 *
Ding Yang, Cao Dun, Wei Wei, Han Zeyang, Su Xiaoming

丁杨，内蒙古工业大学建筑学院，博士研究生。
曹敦，内蒙古工业大学建筑学院，硕士研究生。
魏炜，内蒙古工业大学建筑学院，硕士研究生。
韩则阳，内蒙古恒瑞能源综合管理有限公司，土建工程师。
苏晓明，内蒙古工业大学建筑学院，教授，博士生导师。

摘要 内蒙古地区城镇位于草原腹地，夜间照明干扰小、广域光形态相对独立，具备观察及研究广域光的优势条件。本文通过对锡林浩特市等 9 个典型城镇进行实地调研测试，利用光学测量法及图像分析技术，测量城镇内部的地面照度和星等亮度数据，获取城镇外部广域光侧面图像及剖面灰度值变化曲线等信息。得出地面照度与天顶亮度之间呈指数相关；城镇的内部照明强度、平面形态及内部功能分区是影响广域光形态的主要因素。该研究结果可为城镇夜间光污染的评价、管理及发展预测提供参考依据。

关键词 广域光；草原城镇；夜空发亮；现状调查；分析

Abstract The towns in Inner Mongolia are less disturbed by the nighttime lighting, and the morphology of the wide-area light is relatively independent, which has the conditions for people to investigate. This paper conducts field investigation and test on 9 typical towns such as Xilinhot City. The researchers used optical measurement and image analysis techniques to measure ground illuminance and magnitude luminance data inside the town. The purpose is to obtain the information of the wide area optical profile image and the gray value of the profile outside the town. Conclusion: There is an exponential correlation between ground illuminance and zenith luminance. The interior lighting intensity, plan morphology and internal organization zoning of towns are the main factors affecting the wide-area light form. The results of this study can provide a reference for the evaluation, management and development prediction of urban night light pollution.

Keywords Wide-area light; Prairie town; Night sky light; Status survey; Analysis

---

草原暗夜空保护区域划分研究

Study on the Division of Grassland Dark Night Sky Protection Area

苏晓明，曹敦，丁杨，赵琪，党睿 *
Su Xiaoming, Cao Dun, Ding Yang, Zhao Qi, Dang Rui

苏晓明，内蒙古工业大学建筑学院，教授，博士生导师。
曹敦，内蒙古工业大学，建筑技术科学方向，硕士研究生。
丁杨，内蒙古工业大学，建筑技术科学方向，博士研究生。
赵琪，东营市垦利区住房和城乡建设局，职员。
党睿，天津大学建筑学院，教授，院长助理。

摘要 城市夜间照明正不断蚕食自然夜空环境，影响生态系统。建立暗夜保护区是保护夜行生物栖息地、保护天文观测地的重要举措。草原开阔的视野和清新的空气为夜空观测和建立暗夜保护区提供了良好的条件。本文选取直接影响草原暗夜的城市照明环境开展研究，选取呼伦贝尔草原等进行实地调研与卫星图像数据分析，依据暗夜空分类标准，优化城市照明控制区与控制指标、明确暗夜保护区域范围、提出暗夜保护策略与方法，说明城市照明与草原暗夜保护区域之间的关系，为城市照明规划、自然环境保护研究提供参考。

关键词 草原夜空；保护方法；暗夜公园；城市照明

Abstract Urban nighttime lighting is encroaching on the natural night sky environment and affecting ecosystems. The establishment of the dark night protection area is an important measure to protect the habitats of nocturnal creatures and astronomical observation sites. The open view and fresh air of the grassland provide good conditions for night sky observation and the establishment of dark night protection areas. This paper selects the urban lighting environment that directly affects the object to carry out research. Hulunbuir grassland and other grasslands were selected for field investigation and satellite image data analysis, and the relationship between cities and grassland protection areas was studied from four aspects: classification standard of the dark night sky, lighting control area and control index, protection area definition and protection strategy, and the protection method of grassland dark night sky was proposed. It provides a reference for urban lighting planning and natural environment protection.

Keywords Grassland night sky; Protection method; Dark night park; Urban lighting

建筑立面照明的能耗比较

Comparison of Energy Consumption of Building Facade Lighting

庄金迅，汪江
Zhuang Jinxun,
Wang Jiang

摘要 建筑立面照明是城市夜景照明的一个重要组成部分，其能耗大小对城市夜景照明能耗构成重要影响，因此，实现建筑立面照明节能可以助力我国“双碳”目标的实现。本文依据实际照明工程，统计了媒体立面照明、LED小型投光照明和LED线条灯洗墙照明三种立面照明方式的能耗情况，并对不同照明方式的能耗进行了对比分析，以期对以后类似立面照明的能耗评估分析作参考。

关键词 夜景照明；建筑立面照明；照明能耗；照明方式

Abstract Building facade lighting is an important part of urban night scene lighting, and its energy consumption has an important influence on urban night scene lighting energy consumption, the realization of energy-saving building facade lighting can help achieve our country' s carbon peaking and carbon neutrality goals. According to the actual lighting project, the energy consumption of media facade lighting, LED miniature projection lighting and LED line lamp wall washing lighting are calculated, and the energy consumption of different lighting modes is compared and analyzed in order to provide a reference for the energy consumption evaluation of similar facade lighting in the future.

Keywords Night scene lighting; Building facade lighting; Lighting energy consumption; Lighting mode

庄金迅，辽宁科技大学建筑与艺术设计学院，副教授，教师。
汪江，辽宁科技大学建筑与艺术设计学院，教授，副院长。

基于颜色偏好的博物馆中国绘画照明指标

Indicators for Lighting Chinese Paintings in Museums Based on Color Preference

李卓，王虎，张婧婧，党睿
Li Zhuo, Wang Hu,
Zhang Jingjing, Dang Rui

摘要 博物馆绘画照明对文物光照保护和颜色呈现效果有着双重需求，是最复杂的建筑光环境之一，其中照度（E）和相关色温（CCT）是关键性照明指标。大量研究从保护需求出发提出了E和CCT的阈值范围，但基于颜色偏好的E、CCT指标尚不明晰。针对如何在保护的基础上提升颜色呈现效果，建立满足保护和展示双重需求的照明指标，本研究在1:1搭建的博物馆绘画展厅中，用28种照明工况照射两种典型中国绘画开展颜色偏好评价实验，通过分析得到了E和CCT对颜色偏好的影响规律，并且100~150 lx的照度匹配3500~5000 K的相关色温，是满足保护和颜色偏好双重需求的照明指标范围。

关键词 照度；相关色温；颜色偏好；博物馆照明

Abstract Museum painting lighting has a dual demand for artifact light protection and color rendering effects, and is one of the most complex architectural light environments, where illuminance (E) and correlated color temperature (CCT) are key lighting indicators. A large number of studies have proposed threshold ranges for E and CCT from conservation needs, but the E and CCT indicators based on color preference are not yet clear. To improve the color presentation effect on the basis of conservation and establish lighting indicators that meet the dual needs of conservation and display, 28 lighting conditions were used to irradiate two typical Chinese paintings in a 1:1 built museum painting exhibition hall to carry out color preference evaluation experiments in this study, and obtained the effect law of E and CCT on color preference, and the E of 100~150 lx matched the CCT of 3500~5000 K, which is the range of lighting index to meet the dual needs of conservation and color preference.

Keywords Illuminance; Correlated color temperature; Color preference; Museum lighting

李卓，郑州大学，讲师。
王虎，郑州大学，硕士研究生。
张婧婧，郑州大学，硕士研究生。
党睿，天津大学建筑学院，教授，院长助理。

浅析新形势下夜景照明的可持续发展

Analysis on the Sustainable Development of Nightscape Lighting Under the New Situation

刘小红
Liu Xiaohong

摘要 在坚持以人民为中心的发展思想、“双碳”目标、“城市更新”以及“数字创新”等新形势下，夜景照明行业正面临“百年未有之大变局”。本文通过分析夜景照明行业当前面临的新形势，重点提出了夜景照明在不同发展时期的特点，剖析在当前新形势下夜景照明存在的主要问题，并展望未来夜景照明可持续发展的趋势，供同行指正参考。

关键词 新形势；夜景照明；可持续发展；发展趋势

Abstract Under the new situation of adhering to the people-centered development philosophy, “carbon peaking and carbon neutrality goals”, “urban renewal”, and “digital innovation”, the urban lighting industry is facing “unprecedented changes in a century”. By analyzing the current new situation facing the urban lighting industry, this article focuses on the characteristics of urban lighting in different development periods, analyzes the main problems existing in urban lighting under the current new situation, and looks forward to the trend of sustainable development of urban lighting in the future for peer correction and reference.

Keywords New situation; Nightscape lighting; Sustainable development; Development trend

刘小红，哈尔滨建筑大学 / 中国建筑科学研究院有限公司建科环能科技有限公司，高级工程师。

导光管的显色性研究

Research on Color Rendering of Light Pipes

周幸，欧阳金龙，王春苑，高庆龙
Zhou Xing, Ouyang Jinlong, Wang Chunyuan, Gao Qinglong

摘要 导光管的显色性是衡量其采光质量的重要指标。然而，以往研究多聚焦导光管的光传输效率，未重视其显色性。本文故以两种不同内壁的长径比为 2~10 的导光管为例，系统地研究了导光管的光谱反射率、长径比对显色性的影响。首先利用 TracePro 软件获得导光管透射光的相对光谱，再根据相对光谱计算透射光的 CIE L*a*b* 坐标，最后评价了导光管的显色性。结果表明，宜优先选择光谱反射率分布均匀且平均值较高的导光管，同时尽可能减小长径比。

关键词 导光管；显色性；CIE L*a*b* 颜色空间；TracePro

Abstract The color rendering of light pipes is an important indicator of its light quality. However, previous studies have mostly focused on the light transmission efficiency of light pipes, without paying attention to their color rendering. Therefore, the effect of spectral reflectance and aspect ratio on the color rendering of light pipes was systematically investigated with two different inner wall light pipes with aspect ratios of 2~10 as an example. The relative spectra of transmitted light of the light pipes were firstly obtained by TracePro software, and then the CIE L*a*b* coordinates of transmitted light were calculated according to the relative spectra, and finally the color rendering of the light pipes was evaluated. The results show that it is preferable to select light pipes with uniform spectral reflectance distribution and high average value, and to reduce the aspect ratio as much as possible.

Keywords Light pipes; Color rendering; CIE L*a*b* coordinate; TracePro

周幸，四川大学建筑与环境学院，学生。
欧阳金龙，四川大学建筑与环境学院，教授。
王春苑，四川大学建筑与环境学院，讲师。
高庆龙，四川大学建筑与环境学院，教授，四川大学高性能绿色建筑与低碳城市研究中心主任。

从采光系数到采光能效

From Daylight Factor to Daylight Efficiency

罗涛，薛鹏，高雅春
Luo Tao, Xue Peng, Gao Yachun

摘要 我国政府提出力争 2030 年前实现碳达峰，2060 年前实现碳中和，作为构建绿色低碳循环发展经济的重要目标。天然采光在改善光环境的同时，也是照明行业实现低碳目标的重要技术措施。在我国标准和工程应用中，通常使用采光系数这一静态评价指标；随着技术的发展，UDI、DA 等动态评价指标也相继提出。但这些指标体现的主要是采光效果，不能直接体现节能的效果，而且在工程应用中不够直观，不便于设计人员掌握。为此，本文提出了采光能效的指标，用于定量评价采光节能效果，可作为传统采光指标的补充。

关键词 低碳；绿色建筑；节能；采光照明一体化

Abstract The Chinese government has proposed to achieve a carbon peak before 2030 and carbon neutrality before 2060 as an important goal of building a green and low-carbon circular development economy. Daylighting is also an important technical measure for the lighting industry to achieve low-carbon goals, as well as improve the light environment. In Chinese standards and engineering applications, the daylight factor is usually used as a static evaluation index. With the development of technology, dynamic evaluation indicators such as UDI and DA have also been proposed. However, these indicators mainly reflect the lighting effect, and cannot directly reflect the effect of energy saving. Moreover, it is not intuitive enough in engineering applications and is not easy for designers to grasp. Therefore, this paper puts forward the indicators of daylight efficiency for quantitative evaluation for energy saving, which can be used as a supplement to traditional daylight indicators.

Keywords Low carbon; Green buildings; Energy-saving; Integration for daylight and electric lighting

罗涛，中国建研院建科环能科技有限公司，研究员。
薛鹏，北京工业大学，副教授。
高雅春，中国建研院建科环能科技有限公司，工程师。

中小学教室光环境测量方法研究

Study on Test Method for Luminous Environment of Primary and Secondary School Classroom

罗涛，张滨，梁树英
Luo Tao, Zhang Bin, Liang Shuying

摘要 近年来，儿童青少年视力健康成为全社会关注的焦点问题。各地区政府投入了大量资金进行中小学教室的照明改造，以改善教室光环境，达到近视防控的目的。然而在实施过程中，由于测量方法的不一致，各方对于光环境是否达标，以及实际光环境指标的争议仍时有发生。为此，我们受相关部门委托，开展了中小学教室光环境测量方法的研究并编制了相关标准。在原有照明测量方法的基础上，针对中小学教室的特点，考虑了天然光、人工照明和教学多媒体等因素的共同影响，这对中小学教室光环境提升改造具有一定的指导作用。

关键词 近视防控；眩光；频闪

Abstract In recent years, the vision health of children has become the focus of attention of the whole society. Governments have invested a lot of money in lighting renovation of primary and secondary school classrooms to improve the light environment of classrooms and achieve the purpose of myopia prevention and control. However, during the implementation process, due to the inconsistency of measurement methods, disputes still occur between parties about whether the luminous environment meets the standards and the actual indicators. To this end, we have been commissioned by relevant departments to conduct research on light environment measurement methods and compile relevant standards. According to the characteristics of primary and secondary school classrooms, the joint influence of daylight, artificial lighting and teaching multimedia is considered, which has a certain guiding role in the improvement and transformation of the luminous environment of primary and secondary school classrooms.

Keywords Myopia prevention and control; Glare; Flicker

罗涛，中国建研院建科环能科技有限公司，研究员。
张滨，中国建研院建科环能科技有限公司，高级工程师。
梁树英，重庆大学城规学院，副教授。

街道空间形态对道路照明天空辉光指数影响的模拟分析

Simulation-Based Analysis for the Effects of Street Canyon Configurations on Sky Glow Index of Roadway Lighting

李媛，高雅春，刘小红
Li Yuan, Gao Yachun, Liu Xiaohong

李媛，中国建研院环能科技光环境与照明研究中心，副研究员。
高雅春，建科环能科技有限公司，工程师。
刘小红，中国建筑科学研究院有限公司，高级工程师。

摘要 以满足技术标准要求的同一支路及其照明设施提供的光环境作为基线，通过对平面围合度（PED）和断面围合度（SED）的不同组合的街道空间进行建模和光通量空间分布计算分析，研究街道空间形态与道路照明上射光通造成的相对天空辉光效应——道路照明单位面积天空辉光指数（SGI）的量化关系；还研究了相同 SED 有无退线距离、相同 PED 不同间隙数量、单双侧围合以及单多层围合对 SGI 影响的差别。

关键词 夜间光环境；天空辉光指数；上射光通比；围合度；全局参数

Abstract The quantitative relationship between street canyon configurations (planar enclosure degree PED, cross-sectional enclosure degree SED) and the relative sky glow impacts (Sky Glow Index per unit area of road lighting SGI) produced by the direct- and reflected-upward emissions of road lighting, were researched based on the simulation-based analysis and calculation of the global luminous flux distribution of luminous environment provided by the same roadway lighting installation designed to meet the requirements of technical standards. In addition, the effects of setback with the same SED, gaps number between buildings with the same PED, single-sided and opposite arrangement and surrounding building layout on SGI were also studied.

Keywords Luminous environment at night; Sky glow index; Upward light ratio; Enclosure degree; Global parameters

---

“双碳”背景下基于人眼视觉特征的城市夜景照明光视效能的研究

Research on the Luminous Efficiency of Urban Nightscape Lighting Based on Human Visual Characteristics Under the Background of Carbon Peaking and Carbon Neutrality

陈神飞，林怡 *，姚其，曾宪宪，张兵
Chen Shenfei, Lin Yi, Yao Qi, Zeng Xianxian, Zhang Bing

陈神飞，曾宪宪，张兵，同济大学建筑与城市规划学院，博士研究生。
林怡，同济大学建筑与城市规划学院，教授。
姚其，复旦大学工程与应用技术研究院，副研究员。

摘要 “双碳”目标已成为城市可持续发展的重要命题和挑战。当前城市照明无效光、溢出光等是增加能耗的重要因素，并已上升为社会问题。我们用高光谱仪采集了上海商业区 50 个照明场景，基于人眼“视见、暗视”函数，从光谱层面对城市夜景照明的光视效能分析，提出一种夜景照明光视效能的评价方法。结果表明，自发光光源的光视效能比非自发光高；高光视效能的建筑材质与光源的光谱相似度低。保证夜景观的美学要求，结合建筑材质特性选择优质匹配光源，提高夜景照明光视效能，减少碳排放应是当前设计师的着力点。

关键词 光视效能；城市照明；能耗评估

Abstract The pursuit of carbon peaking and carbon neutrality goals has emerged as a crucial goal and challenge for sustainable urban development. Ineffective lighting and light pollution are key factors contributing to excessive energy consumption, thereby becoming a pressing social issue. This study collected 50 hyperspectral images of lighting scenes in the Shanghai business district. With consideration of the photopic and scotopic functions of the human eye, we analyzed the luminous efficiency of urban nightscape lighting at a spectral level and proposed an evaluation method for such lighting. Results demonstrate that the luminous efficacy of self-luminous light sources is higher than that of non-self-luminous ones, and building materials with high luminous efficacy have low spectral similarity to the light source. To ensure the aesthetic requirements of night scenes and enhance luminous efficacy while reducing carbon emissions, it is crucial for designers to carefully select high-quality and matched light sources based on the characteristics of building materials.

Keywords Luminous efficiency; Urban lighting; Energy consumption assessment

与天然光联动的理想人工照明系统能耗估算

Estimating the Energy Consumption of an Ideal Daylight-Linked Artificial Lighting System

遇大兴，张祖玮，冯丽明，边宇
Yu Daxing, Zhang Zuwei, Feng Liming, Bian Yu

遇大兴，华南理工大学建筑学院，副教授。
张祖玮，华南理工大学建筑学院，硕士研究生。
冯丽明，华南理工大学建筑学院，硕士研究生。
边宇，华南理工大学建筑学院，副教授。

摘要 本研究旨在探讨有采光房间的日间照明能耗问题，基于房间采光情况预测与天然光联动的室内照明控制系统运行能耗。文中提出并使用了“天然光缺乏量”（IDQ）这一概念用于描述某空间采光不足程度。通过理论分析与数学建模，基于某空间的IDQ和照明方案的利用系数得出了照明能耗计算模型。在此基础上，在指定的环境条件下，分析了房间参数与照明能耗的数量关系。最终，基于多样本计算结果获得了日间照明能耗与SDA的数量关系，该结果可推广到不同气候条件、不同参数的房间的应用场景，用于预测与天然光联动的室内照明系统能耗。

关键词 采光；照明；天然光联动控制；能耗预测

Abstract This study investigates the daytime lighting energy consumption in lighted rooms and to predict the energy consumption of an indoor lighting control system operating in conjunction with natural light based on room lighting conditions. The concept of “natural light deficiency quantity (IDQ)” is proposed and used to describe the degree of light deficiency in space. Through theoretical analysis and mathematical modeling, a lighting energy consumption calculation model is derived based on the IDQ of space and the utilization factor of a lighting scheme. On this basis, the quantitative relationship between room parameters and lighting energy consumption was analyzed under specified environmental conditions. Finally, the quantitative relationship between daytime lighting energy consumption and SDA was obtained based on the results of the multi-sample calculation, which can be extended to application scenarios of rooms with different climatic conditions and parameters for predicting the energy consumption of indoor lighting systems linked with natural light.

Keywords Daylight; Lighting; Daylight linked control; Energy consumption prediction

---

中医体质作为个体光敏性差异识别方法的研究与探索

Research and Exploration of Traditional Chinese Medicine Constitution as a Method to Identify Individual Photosensitivity Differences

翟会芳，严永红 *
Zhai Huifang, Yan Yonghong

翟会芳，重庆大学建筑城规学院，博士研究生。
严永红，重庆大学建筑城规学院，教授。

摘要 当前研究已证实，人体对光的视觉和非视觉效应有显著的个体差异。在群体中快速识别出差异性，将助于个性化、精准化健康照明的发展。本文基于电竞青训运动员日常训练的实地研究，以中医体质分型理论为切入点，对同一训练光环境下，电竞运动员全天作业下的视疲劳、KSS和身体疲劳3项指标进行分析，探索中医体质作为个体光敏性差异识别方法的可行性。结果表明，在同一训练光环境下，平和质与偏颇体质的运动员相比，在全天候训练下的整体视疲劳、KSS和身体疲劳均存在显著差异。

关键词 中医体质；个体差异；光敏性；健康照明；电子竞技

Abstract Current research has confirmed that there are significant individual differences in the visual and non-visual effects of light in humans. To quickly identify the differences can facilitate the development of personalized and precise health lighting. In this paper, based on the field study of professional e-sports athletes' daily training and taking classification and determination of constitution in Traditional Chinese medicine(TCM) as a starting point, three indices of visual fatigue, KSS, and physical fatigue were analyzed for e-sports athletes under the same training light environment with all-day work to explore the feasibility of TCM physique as a method to identify individual photosensitivity differences. The results showed that there were significant differences in overall visual fatigue, KSS, and physical fatigue between athletes in the balanced constitution compared to those in the unbalanced constitutions under the same training light environment and under all-weather training.

Keywords Traditional Chinese medicine (TCM) constitution; Individual differences; Photosensitivity; Healthy Lighting; E-sports

基于气候的采光模型的建筑天然光环境研究进展

Research Progress of Building Natural Light Environment Based on Climate-Based Daylight Modeling

梁树英，杨易
Liang Shuying, Yang Yi

摘要 天然光作为绿色能源，对建筑可持续发展有着重要意义。正确评估建筑自然采光是积极利用天然光的关键因素，而目前的天然采光研究主要以静态计算评估为基础，未考虑到天然光的高度动态性以及地域不同对室内采光的影响。因此，基于气候的采光模型（CBDM）的提出对建筑天然采光的适用性应用提出了新的要求。本文回顾了近十几年来国内外对 CBDM 的研究成果，从“量化研究”“技术应用”“度量标准”三个方面进行了梳理，总结了基于 CBDM 建筑天然采光的研究进展及其未来发展趋势。

关键词 天然采光；基于气候的采光模型（CBDM）；光气候数据；天空亮度模型；动态采光

Abstract As a green energy source, daylight has great significance for the sustainable development of architecture. Correct evaluation of building daylighting is a key factor in the active use of daylight, while current research is mainly based on static calculation assessment, without taking into account the highly dynamic of daylight and the influence of geographic differences on indoor daylighting. Therefore, the proposal of Climate-based Daylight Modeling(CBDM) has put forward new requirements for the applicability application of building daylighting. This paper reviews the research results on CBDM at home and abroad in the past ten years, sorts out the research results from quantitative research, simulation technology and metric standard, summarizes the research progress of building daylighting based on CBDM, and proposes the future development.

Keywords Daylighting; Climate-based Daylight Modeling; Daylight climate data; Sky brightness model; Dynamic daylighting

梁树英，重庆大学建筑城规学院，副教授，硕士生导师，主要从事建筑技术科学研究。
杨易，重庆大学建筑城规学院，建筑学硕士。

建筑照明逆设计生成与多目标优化的软件实现

Software Implementation of Inverse Design Generation and Multi-objective Optimization for Architectural Lighting

陈鹏，王立雄，吴雨婷，于娟*，王爱英，孔光燕
Chen Peng, Wang Lixiong, Wu Yuting, Yu Juan, Wang Aiying, Kong Guangyan

摘要 为应对照明与建筑专业设计流程协同、目标协调的设计集成新策略，引入了生成式设计手段，将照明设计过程转化为对照明指标的优化过程。明确了照明逆设计与多目标优化的适用启发式算法，确定了 Revit 是照明建筑设计集成的实施平台。重构了照明设计流程，针对“环境质量提升、空间功能强化、照明节能优化”新目标，提炼出核心照明指标和设计要素，明确了要素和指标在优化中的地位、逆设计与多目标优化方法的选择依据。编写了算法程序实施流程，成果有助于提高设计质量和效率、推动照明建筑一体化深入发展。

关键词 生成式设计；建筑照明；逆设计；多目标优化；启发式算法

Abstract To cope with the new design integration strategy of the synergistic process and goal coordination between lighting and architectural design, generative design means are introduced to transform the lighting design process into an optimization process for lighting metrics. The applicable heuristic algorithms for inverse lighting design/multi-objective optimization are clarified, and Revit is identified as the implementation platform. The lighting design process is reconstructed, and core lighting indicators and design elements are extracted under the new objectives of "quality improvement, function enhancement and energy saving". The role of the elements and indicators in the optimization, and the principle for the choice between the inverse design/multi-objective method, are clarified. Software was developed to implement the process, and the outcome will help improve the design quality and efficiency, and promote the further development of the integration of building and lighting.

Keywords Generative design; Architectural lighting; Inverse design; Multi-objective optimization; Heuristic algorithm

陈鹏，天津大学，博士研究生。
王立雄，天津大学，教授。
吴雨婷，天津大学，博士研究生。
于娟，天津大学，助理研究员。
王爱英，天津大学，副教授。
孔光燕，天津大学，博士研究生。

天然采光技术在绿色建筑中的应用

Application of Natural Lighting Technology in Green Buildings

李烨，张明宇
Li Ye, Zhang Mingyu

摘要 天然采光作为绿色建筑设计的重要技术手段，可以提高视觉舒适性，降低建筑照明能耗，进而改善居住或工作环境。文章介绍了天然采光研究方法、内容等发展现状，梳理了绿色建筑采光设计的要求、方法和影响要素等，并结合具体案例说明天然采光技术在绿色建筑中的应用效果，以期为今后绿色建筑相关工作提供参考。

关键词 天然采光；绿色建筑；发展现状；采光技术

Abstract As an important technical means of green building design, natural lighting can improve visual comfort, reduce the energy consumption of building lighting, and thus improve the living or working environment. The article introduces the development status of natural lighting research methods and contents, sorts out the requirements, methods and influencing elements of green buildings lighting design, and illustrates the application effect of natural lighting technology in green building with specific cases, in order to provide reference for future green building related work.

Keywords Natural lighting; Green building; Development status; Daylighting technology

李烨，天津大学建筑学院，博士研究生。
张明宇，天津大学建筑学院，副教授，建筑技术科学研究所副所长。

狭小封闭空间中有无视野的人工窗口的恢复作用比较

Comparison of the Restorative Effect of Artificial Window Between with and without View in Isolated and Confined Space

唐博，张心宇，张昕 *
Tang Bo, Zhang Xinyu, Zhang Xin

摘要 狭小封闭空间隔绝天然光的同时也隔绝了视野，这会造成人员缺少与外界接触，影响人员生理心理健康状态。地上空间研究表明，通过引入节律照明和视野图像可以改善人员生理心理健康状态。本研究关注视野与节律照明，在动态照明条件下设计人工窗口，收集生理数据和心理主观评价并比较其在狭小封闭空间中的恢复作用。研究结果显示，人工窗口的恢复性作用明显。视野对窗口恢复作用的影响不显著。研究结果补充了狭小封闭空间中视野与动态照明的比较分析，为狭小封闭空间人工窗口设计和评估提供参考。

关键词 狭小封闭空间；人工窗口设计；节律照明；视野；人因

Abstract People in isolated and confined space are separated from both natural light and window view, which results in a lack of contact with the outside world and affects the physiological and psychological health of personnel. Research on aboveground space has shown that introducing dynamic circadian lighting and window view can improve the overall physiological and psychological health of personnel. This study focuses on the window view and circadian lighting and designs artificial window under dynamic circadian lighting conditions. Physiological data and psychological subjective evaluation are collected to compare the restorative effect of the artificial window in isolated and confined spaces. The results show that the restorative effect of dynamic lighting on artificial windows is significant. The effect of view on window' s restoration is not significant. The results supplement the comparative analysis of view and dynamic lighting, providing reference for the design and evaluation of artificial windows in isolated and confined space.

Keywords Isolated and confined space; Artificial window design; Circadian lighting; Window view; Human factors

唐博，清华大学建筑学院，博士研究生。
张心宇，清华大学建筑学院，本科生。
张昕，清华大学建筑学院，副教授。

基于动作幅度检测的室内工作面照明控制研究

Research on Indoor Working Face Lighting Control Based on Motion Amplitude Detection

王智德，张明宇，张孟源，李升东，于晓晴

Wang Zhide, Zhang Mingyu, Zhang Mengyuan, Li Shengdong, Yu Xiaoqing

摘要 本文针对现有主流照明控制方式灵敏度较低、智能化较低且无法真正自主做出控制决策的问题，提出了一种新型灯光控制方法，即通过对人肢体动作幅度的探测来控制灯光。文章首先回顾了相关领域的研究，包括肢体动作探测和灯光控制技术的发展。其次以厨房炊事活动为例，通过实验得出了人体在不同室内光环境下肢体动作的变化规律，确定人在推荐工作面照度值下工作时的肢体姿态。最终对实验数据进行整理和分析后，进行控制系统软件的开发设计，提升灯光控制的智能化以及人性化程度。

关键词 动作检测；室内照明；工作面照度；照明控制

Abstract In this paper, aiming at the problems of low sensitivity, low intelligence and inability to make control decisions independently of the existing mainstream lighting control methods, a new lighting control method is proposed, that is, to control lights by detecting the amplitude of human body movements. The paper firstly begins with a review of research in related fields, including the development of body motion detection and light control techniques. Secondly, taking cooking activities in the kitchen as an example, the changing rules of body movements in different indoor light environments are obtained through experiments, and the body posture of people working under the recommended working face illuminance value is determined. Finally, after sorting out and analyzing the experimental data, the control system software was developed and designed to improve the intellectualization and humanization of lighting control.

Keywords Motion detection; Interior lighting; Working face illuminance; Lighting control

王智德，天津大学建筑学院，博士研究生。
张明宇，天津大学建筑学院，副教授，博士生导师。
张孟源，李升东，于晓晴，天津大学建筑学院，硕士研究生。

绘画文物中颜料的光谱响应规律评估

Evaluation of Spectral Response Rule of the Pigment in Painting Relics

谭慧姣，党睿 *

Tan Huijiao, Dang Rui

摘要 在博物馆中 LED 推广应用的背景下，探讨绘画文物中颜料中的光谱响应规律是对其开展照明保护的关键。然而，常用的色差评估方法存在局限性。选定绘画文物中典型的朱砂颜料样品在 9 种窄带 LED 辐照下开展加速光老化实验，照射前后多次测试样品的反射光谱和三刺激值，分别结合主成分分析和 CIE 计算公式进行数据分析，并进一步分析样品在不同窄带 LED 辐照下的照明损伤规律，得到了其光谱响应函数 $f(\lambda)$，即照明损伤与辐照波长 λ 的数学关系，为相关的绘画文物照明保护研究提供理论基础。

关键词 绘画文物；颜料；光谱响应规律；反射光谱；主成分分析

Abstract Under the background of the promotion and application of LEDs in the museum, it is the key to carrying out protective lighting to explore the spectral response of pigments in painting relics. However, there are limitations to the commonly used evaluation method of color difference. The typical cinnabar pigment in painting relics was selected as the sample to carry out an accelerated photoaging experiment under the irradiation of 9 narrow-band LEDs. The reflection spectrum and tristimulus value of samples were tested several times before and after irradiation, which was analyzed by the principal component analysis and CIE calculation formula, respectively. The lighting damage rule of the sample under different narrowband LEDs was further analyzed. The obtained spectral response function $f(\lambda)$, that is, the mathematical relationship between lighting damage and spectral wavelength λ, provides a theoretical basis for related research on lighting protection of painting relics.

Keywords Painting relic; Pigment; Spectral response rule; Reflection spectrum; Principal component analysis

谭慧姣，天津大学，博士。
党睿，天津大学建筑学院，教授，院长助理。

基于视野亮度分布的博物馆展厅眩光量化

Glare Quantification in Museum Exhibition Halls Based on Visual Field Luminance Distribution

高子昂，胡振宇，丛阳，孙星灿，党睿
Gao Ziang, Hu Zhenyu, Cong Yang, Sun Xingcan, Dang Rui

高子昂，天津大学建筑学院，博士研究生。
胡振宇，天津大学建筑学院，博士研究生。
丛阳，天津大学建筑学院，硕士研究生。
孙星灿，天津大学建筑学院，硕士研究生。
党睿，天津大学建筑学院，教授，院长助理。

摘要 博物馆展厅由于复杂的空间亮度分布极易形成强烈的眩光，致使展品观赏所需的视觉效果难以实现。通过建立眩光计算模型从而对眩光进行精确的量化评估，是改善眩光问题的理论基础。本文首先在全尺寸可变空间实验舱中复现了真实的博物馆展厅，进而在确保藏品光安全和显色效果的前提下通过改变环境照明参数和人员视看点位参数获得了 288 种亮度分布条件，一方面基于视野亮度分布测试出眩光模型 UGR 的准确率为 41.67%；另一方面基于眩光评价实验结果构建了新的眩光模型 MGR，经验证其准确率可达 78.13%，能够实现博物馆展厅眩光的量化计算及等级评价。

关键词 博物馆展厅；视野亮度分布；眩光计算模型；眩光等级评价

Abstract The complex spatial luminance distribution of museum exhibition halls makes it difficult to achieve the visual effects required for exhibit viewing because of the strong glare. An accurate quantitative assessment of glare through the establishment of a glare calculation model is the theoretical basis for improving the glare problem. In this study, we first reproduced a real museum exhibition hall in a full-size variable space experiment chamber, and then obtained 288 luminance distribution conditions by changing ambient lighting parameters and personnel viewing point parameters while ensuring the light safety and color rendering effect of the collection. On the other hand, a new glare model MGR was constructed based on the experimental results of glare evaluation, and its accuracy was verified to be 78.13%, which can realize the quantitative calculation and level evaluation of glare in museum exhibition halls.

Keywords Museum exhibition halls; Visual field luminance distribution; Glare calculation model; Glare rating evaluation

---

基于亮度梯度的教室照明视觉疲劳预测模型

Visual Fatigue Prediction Model for Classroom Lighting Based on Luminance Gradients

刘宇星，常锁，党睿
Liu Yuxing, Chang Suo, Dang Rui

刘宇星，天津大学建筑学院，博士研究生。
常锁，天津大学国际工程师学院，硕士。
党睿，天津大学建筑学院，教授，院长助理。

摘要 教室照明的质量对学生的学习效率和眼睛健康有重要影响。为了降低视觉疲劳，需要设置合理的照明参数组合。然而，常用的照度和色温值无法全面反映光环境的复杂性和非视觉效应对生理变化的影响。因此，本研究引入了亮度梯度概念，并提出了教室照明的 20 个光环境特征值。通过主观疲劳感评分和闪光融合频率测试，发现 4 个照明参数对视觉疲劳的影响显著。研究结果为评价教室照明质量和设计健康教室提供了量化工具，建立了视觉疲劳的综合指标预测模型 $VF=g(D_{31}, D_{12}, L_m, CCT)$。

关键词 教室照明；亮度梯度；相关色温；视觉疲劳；数学模型

Abstract The quality of classroom lighting has a significant impact on students' learning efficiency and eye health. To reduce visual fatigue, it is necessary to set reasonable combinations of lighting parameters. However, commonly used values for illuminance and correlated color temperature cannot comprehensively reflect the complexity of the light environment and the effects of non-visual factors on physiological changes. Therefore, this study introduces the concept of luminance gradients and proposes 20 light environment parameters for classroom lighting. Through subjective fatigue ratings and flicker fusion frequency tests, it was found that four lighting parameters significantly affect visual fatigue. The results of this study provide a quantitative tool for evaluating classroom lighting quality and designing healthy classrooms. A prediction model for the comprehensive index of visual fatigue, $VF=g(D_{31}, D_{12}, L_m, CCT)$, has been established.

Keywords Classroom lighting; Luminance gradient; Correlated color temperature; Visual fatigue; Mathematical model

中国彩绘文物照明：保护性能增强的光谱可调型 WLEDs 光源

Lighting Chinese Painted Cultural Relics: Emerging Spectrally Tunable WLEDs Light Source System with Enhanced Protection Performance

党睿*，胡振宇，蔡响前
Dang Rui, Hu Zhenyu, Cai Xiangqian

党睿，天津大学建筑学院，教授，院长助理。
胡振宇，天津大学建筑学院，博士研究生。
蔡响前，天津大学国际工程师学院，硕士研究生。

摘要 在中国彩绘文物照明中，要求光源不仅具有高质量的展示效果，还需尽量降低对文物造成的光损伤。本文通过实验方法提出了在照度、时间、光源光谱、材料的光谱响应率耦合影响下文物光损伤的预测模型，平均误差为9.37%。将该模型与既有展示性能计算方法结合建立了光谱优化算法，并与10 通道 LED 系统结合制备出新型 WLEDs 光源系统，该系统可根据文物材料特点调节并输出与之匹配的最优光谱，利用一套系统在满足观赏的条件下实现不同文物的最低损伤照明。

关键词 中国彩绘文物；光损伤预测模型；光谱优化算法；光谱可调型 WLEDs

Abstract When lighting Chinese-painted cultural relics, the light sources are required not only to provide high-quality visual performance, but also to minimize the photodamage. In this study, a photodamage prediction model with an error of 9.37% predicting the photodamage degree of CPCRs under the coupling influence of four parameters, namely, illuminance, time, relative spectral power distribution (SPD) and spectral responsivity of materials, was established. Combined with the existing methods for calculating the visual performance indexes, a spectral optimization algorithm was developed, followed by the development of a new Spectrally Tunable White LEDs (STWLEDs) light source system with 10 LED chips. It is found that the STWLEDs can adjust and output the optimal spectral according to the characteristics of CPCRs, thus realizing lighting with minimum photodamage, meanwhile, satisfying high-quality visual performance.

Keywords Chinese painted cultural relics; Photodamage prediction model; Spectral optimization algorithm; Spectrally tunable WLEDs

---

教室照明中亮度参数与相关色温对视觉功效影响研究

A Study on the Impact of Luminance Parameters and Correlated Color Temperature on Visual Performance in Classroom Lighting

赵芮雪，刘宇星，常锁，党睿*
Zhao Ruixue, Liu Yuxing, Chang Suo, Dang Rui

赵芮雪，天津大学建筑学院，硕士研究生。
刘宇星，天津大学建筑学院，博士研究生。
常锁，天津大学国际工程师学院，硕士。
党睿，天津大学建筑学院，教授，院长助理。

摘要 教室的夜间使用越来越频繁，教室照明对创造学习氛围和提高学习效率的作用愈发重要。传统照明标准忽略了照明环境的协调统一性和与人的视觉感受的联系。本文研究还原了教室照明场景，引入更全面、与人相关和更能反映教室环境特征的亮度参数，辅以色温参数，建立了包含 20 项参数的教室照明环境衡量体系。通过主观评价和客观视觉任务测试实验，采用最大互信息系数分析变量参数与主客观评价指标之间的关系，并利用机器学习算法建立了基于视觉功效的教室照明环境评价模型，为教室照明参数优化提供科学依据。

关键词 教室照明；亮度参数；视觉功效；评价模型；相关色温

Abstract Classrooms are used more and more frequently at night, so classroom lighting plays an increasingly important role in creating learning atmosphere and improving a learning efficiency. Traditional lighting standards ignore the coordination and unity of the lighting environment and the connection with human visual perception. This study restored the classroom lighting scene, introduced more comprehensive, human-related and more reflective brightness parameters of the classroom environment characteristics, supplemented by color temperature parameters, and established a classroom lighting environment measurement system containing 20 parameters. Through subjective evaluation and objective visual task testing experiments, and using the maximum mutual information coefficient to analyze the relationship between the variable parameters and subjective and objective evaluation indicators. The evaluation model of classroom lighting environment based on visual efficiency is established by using a machine learning algorithm, which provides the scientific basis for the optimization of classroom lighting parameters.

Keywords Classroom lighting; Luminance parameter; Visual effect; Evaluation model; Related color temperature

光源色彩参数对中国绘画颜色偏好的影响规律及其数学表征

The Influence Law of Light Source Color Parameters on the Color Preference of Chinese Paintings and Its Mathematical Representation

王舒宁，孙星灿，唐铭，杨頔，党睿*
Wang Shuning, Sun Xingcan, Tang Ming, Yang Di, Dang Rui

王舒宁，孙星灿，杨頔，天津大学建筑学院，硕士。
唐铭，中国国家博物馆，副高级，文保院副研究馆员。
党睿，天津大学建筑学院，教授，院长助理。

摘要 中国画对于光源有保护和展示的双重要求。保护需求决定了展厅光环境低照度和低色温的特点，因此如何在这种光环境下针对中国画特殊颜色进行颜色偏好的量化评价是亟待解决的问题。本文通过在搭建的真实博物馆展厅中开展多照明参数耦合下的中国画色彩偏好度心理物理学评价实验，进而通过相关性分析等方法，得到了光源色彩参数对中国画颜色偏好的影响规律和颜色偏好评价方程，可为中国画照明中的光源比较与选择、该类展品的专用光源开发与设计提供数学工具。

关键词 博物馆照明；中国绘画；光源色彩参数；颜色偏好；数学模型

Abstract There are two requirements for light sources in the protection and display of Chinese paintings. Considering the lighting environment in an exhibition with low illuminance and color temperature determined by the protection requirement, how to quantitatively evaluate the color preference for the special colors of Chinese paintings remains a challenge. In this study, an influence law of light source color parameters on the color preference of Chinese paintings is proposed through a psychophysical evaluation experiment under the coupling of multiple lighting parameters and conducted in a real museum exhibition hall, as well as using several analysis methods of correlation analysis, etc. The color preference of Chinese paintings could then be evaluated. In this way, mathematical solutions for the comparison and selection of light sources, as well as their development and design, for Chinese paintings are proposed.

Keywords Museum lighting; Chinese painting; Light sources' color parameters; Color preference; Mathematical model

---

中国绘画照明光源的颜色质量评价方法及系统

Color Quality Evaluation Method and System of Chinese Painting Lighting Source

谷高利，孙星灿，党睿
Gu Gaoli, Sun Xingcan, Dang Rui

谷高利，天津大学未来技术学院，硕士研究生。
孙星灿，天津大学建筑学院，硕士研究生。
党睿，天津大学建筑学院，教授，院长助理。

摘要 如何在展示过程提升绘画的观赏质量是让“文物活起来”至关重要的一步，但中国绘画在照明环境和自身颜色两方面具有特殊性，因而现有的颜色质量评价方法并不适用于中国绘画。本文通过开展对中国绘画色彩感知心理物理实验，通过皮尔逊相关性及主成分分析法，得到以 IES R_f、IES R_g 和 IES $R_{cs,h5}$ 三个指标为关键影响因子的色彩真实度和喜好度评价模型，并给出了指标阈值推荐。同时，以评价模型为核心，开发了一套中国绘画照明的光源颜色质量测量及评价系统，为绘画照明光源的选择、各类新型光源准入判定提供了科学的手段和依据。

关键词 颜色质量；博物馆光源；中国绘画；光源评价

Abstract How to improve the viewing quality of paintings in the process of the display is a crucial step to make "cultural relics alive". However, because of the particularity of Chinese paintings in the lighting environment and their color, the existing color quality evaluation methods are not suitable for Chinese paintings. By conducting psychophysical experiments on color perception in Chinese paintings, this study obtained a color authenticity and preference evaluation model with IES R_f, IES R_g and IES $R_{cs,h5}$ as the key influencing factors through Pearson correlation and principal component analysis, and recommended the threshold values of the indicators. At the same time, with the evaluation model as the core, a set of color quality measurement and evaluation system of Chinese painting lighting source is developed, which provides scientific means and basis for the selection of painting lighting source and the admission decision of various new light sources.

Keywords Color quality; Museum light source; Chinese painting; Light source evaluation

高校建筑学专业教室光环境浅析

Analysis on Light Environment of Classrooms for Architectural Specialty High Education

幸倩如，傅秀章
Xing Qianru, Fu Xiuzhang

摘要 教室光环境对学生的身心健康、学习效率等有重要影响。高校建筑学专业教室内有手绘、模型制作、电脑制图和课堂教学等不同类型的教学活动，是建筑学专业学生的主要学习场所。专业教室光环境应匹配不同的教学活动。本文通过对国内外的高校教室光环境规范标准比较研究，分析了国内外在建筑学专业教室光环境要求方面的特点与差异，发现目前国内的规范标准未结合不同场景进行特殊要求，天然采光缺少动态分析，建筑学专业教室缺少针对性的光环境设计要求。最后，分别以手绘、模型制作、电脑制图和课堂教学为研究对象，提出了若干光环境要求与建议。

关键词 专业教室；光环境；教学活动；指标

Abstract The classroom light environment has an important impact on students' learning efficiency and physical and mental health. The architectural classroom in universities have different types of teaching activities such as hand drawing, model making, computer graphics and classroom teaching, which are the main learning places for architectural students. The light environment of architectural classrooms should match the different teaching activities. This paper analyzed the characteristics and differences between domestic and international requirements for the light environment of architectural classrooms through a comparative study of domestic and international norms and evaluation standards for the light environment of university classrooms. Finally, four types of teaching activities were used as research objects to propose a number of light environment requirements and suggestions, including hand drawing, model making, computer graphics and classroom teaching.

Keywords Architectural classroom; Light environment; Teaching activities; Criteria

幸倩如，东南大学建筑学院，硕士研究生。
傅秀章，东南大学建筑学院，副教授，东南大学建筑物理实验室主任、建筑物理与设备研究所所长。

博物馆展厅的光色偏好研究

Research on Light-color Preference in a Museum Exhibition Hall

李小凡，朱上宇，黄险峰 *，吴楷文，王一辰，冯俊恒
Li Xiaofan, Zhu Shangyu, Huang Xianfeng, Wu Kaiwen, Wang Yichen, Feng Junheng

摘要 照明是营造博物馆视觉体验的重要因素，也是影响其艺术氛围的关键元素。本文对博物馆常见照明方式进行归纳，从色温和显色性两个方面的评价来分析光环境质量，进而探寻观赏者对不同光敏感性展品的色温和显色性的偏好。结果表明：2900 K、3000 K、4000 K 分别在对光特别敏感、较敏感以及不敏感性的展品中的色温评价较好。随后，对显色值（Ra）为 80、90、97 的三种 LED 光环境情况进行对比分析，发现 Ra 为 97 的照明环境更具有吸引力。这些可为博物馆光环境设计提供指导。

关键词 博物馆展厅；光色偏好；色温；显色指数

Abstract Artificial lighting is one of the major factors affecting the visual experience in museum exhibition hall, and is also a key element to create the artistic atmosphere of the museum. This paper summarizes the common lighting methods in museums, and analyzes the light environment quality from two aspects: the preference of color temperature and the evaluation of color rendering. By the evaluation experiment, the viewers preferences on the color temperature and color rendering under different conditions were investigated. The results show that the color temperature evaluation of 2900 K, 3000 K and 4000 K is better in the exhibits that are particularly sensitive to light, sensitive and insensitive. Subsequently, the environments under three LED light sources with color rendering index (Ra) of 80, 90 and 97 were comparative analyzed, and the lighting environment with color rendering values (Ra) of 97 was the most attractive. This study will be beneficial to the optical environment design of the museum exhibition hall.

Keywords Museum exhibition hall; Light color preference; Correlated color temperature; Color rendering index

李小凡，广西大学土木建筑工程学院，硕士研究生。
朱上宇，广西大学毕业，硕士研究生。
黄险峰，广西大学土木建筑工程学院，教授。
吴楷文，广西大学土木建筑工程学院，硕士研究生。
王一辰，冯俊恒，广西大学土木建筑工程学院，本科生。

“双碳”背景下的乡村照明建设思考

Reflections on Rural Lighting Design Based on the Background of Carbon Peaking and Carbon Neutrality

刘小红，李媛
Liu Xiaohong, Li Yuan

摘要 随着全球能源需求的不断增长，以及对环境保护的日益关注，“双碳”经济已成为未来发展的重要方向，乡村照明建设在“双碳”时代背景下也面临着新的挑战和机遇。本文通过分析当前乡村照明建设的现状，探讨了在“双碳”背景下乡村照明建设的原则、标准和具体方法，旨在为乡村照明可持续发展提供有益的参考。

关键词 乡村照明；“双碳”；可持续发展

Abstract With the continuous growth of global energy demand and increasing attention to environmental protection, the economy of carbon peaking and carbon neutrality has become an important direction for future development. Rural lighting construction is also facing new challenges and opportunities in the context of the “carbon peaking and carbon neutrality” era. This paper analyzes the current situation of rural lighting construction and explores the principles, standards, and specific methods of rural lighting construction under the background of carbon peaking and carbon neutrality, aiming to provide useful references for the sustainable development of rural lighting.

Keywords Rural lighting; Carbon peaking and carbon neutrality; Sustainable development

刘小红，哈尔滨建筑大学，高级工程师。
李媛，天津大学，高级工程师。

南极科考智能人因健康支持系统研究

Research on Intelligent Human Factors Health Support System for Antarctic Scientific Research

郝洛西，于航，靳令经，刘伟静，魏力，王安
Hao Luoxi, Yu Hang, Jin Lingjing, Liu Weijing, Wei Li, Wang An

摘要 南极低氧高寒、极昼极夜、隔绝的恶劣环境对科考人员身心健康造成多重严峻挑战。围绕如何整合建成环境要素，为科考队员提供健康保障，降低极地综合征、情感障碍的发生风险，团队完成了南极健康人因疗愈舱的研发。疗愈舱集成了智能照明系统、环境监测系统、站区噪声掩蔽系统、声景疗愈、疗愈绿植墙等多项自主研发技术，可实现环境精准调控与多维度健康干预。本文从极地环境与人体健康关联机理、建成环境主动健康关键技术到集成解决方案，系统性地介绍了疗愈舱“理、工、医”交叉与“产、学、研”协作的研发路线及其整合的创新健康人居技术。

关键词 极地科考；人因；健康支持；环境疗愈

Abstract The Antarctic’s low oxygen, high cold, extreme day and night, and isolation challenge the scientific research crew’s physical and mental health. The Antarctic Health Human Factors Healing Module, which integrates built environment elements to protect research crews from polar syndrome and emotional disorders, was completed. The healing module uses self-developed technologies like intelligent lighting, environmental monitoring, station noise masking, soundscape healing, and healing greenery wall for precise environmental control and multi-dimensional health intervention. This paper systematically introduces the Healing Module’s research and development route from the intersection of science, engineering, and medicine and the collaboration between industry, academia, and research, to the mechanism of the connection between the polar environment and human health, the key technologies of active health in the built environment, the integrated solutions, and the innovative healthy habitat technologies it integrates.

Keywords Antarctic expedition; Human factor; Health-support; Environmental healing

郝洛西，同济大学建筑与城市规划学院，教授，国际照明委员会副主席。
于航，同济大学建筑与城市规划学院，教授。
靳令经，上海市养志康复医院（上海市阳光康复中心），教授，院长。
刘伟静，上海市第十人民医院，教授，主任。
魏力，中国极地研究中心（中国极地研究所），工程师。
王安，上海亚明照明有限公司，高级工程师，总监。

视黑素照度指标用于节律照明度量的准确性研究

The Accuracy Study of Melanopic Illuminance Metric for Circadian Efficiency Prediction

黄滢滢，戴奇
Huang Yingying, Dai Qi

摘要 当前，基于视黑素照度的节律效应量化方法在健康照明领域中被广泛应用，包括等效视黑素勒克斯（EML）和等效视黑素 D65 照度（m-EDI）。近期研究表明，S 形视锥体也参与了光的节律效应，因此，视黑素照度相关指标的准确性有待进一步探究。本研究采用一对视黑素照度相同、S 形视锥体照度存在较大差异的光谱进行人因实验，发现在较高 S 形视锥体照度条件下，受试者褪黑激素浓度和睡眠效率更高。结果表明，基于视黑素照度的节律效应量化方法准确度不足，应结合 S 形视锥体照度进行优化。

关键词 健康照明；节律效应；视黑素照度；褪黑素抑制；睡眠

Abstract In this circadian phototransduction processes, the melanopsin-containing intrinsically photosensitive retinal ganglion cells (ipRGCs) is the major contributor. Currently, based on this mechanism, there' re some leading models adopt melanopic illuminance for predicting the light induced circadian effects, such as the equivalent melanopic lux (EML) and melanopic equivalent daylight (D65) illuminance model (m-EDI). While, it was found that S-cone visual photoreceptors also contribute to the circadian pathway. Therefore, the accuracy of the widely accepted melanopic illuminance models is urgent to be investigated. In this study, we designed a pair of general-lighting spectra to achieve equivalent melanopic and photopic illuminance, but very different cyanopic illuminance between the spectra-in-contrast. Human-factor studies were performed, and found that the saliva melatonin concentration and sleep efficiency were significantly higher under the lighting condition with the spectrum of greater cyanopic illuminance. Therefore, for an accurate qualification in circadian lighting design, the widely-adopted circadian illuminance metrics that solely based on melanopic illuminance should be modified by combining the contribution of cyanopic illuminance.

Keywords Circadian lighting; Circadian effect; Melanopic illuminance; Melatonin suppression; Sleep

黄滢滢，复旦大学工程与应用技术研究院，博士研究生。
戴奇，复旦大学工程信息科学与工程学院，研究员。

光谱选择性窗户的室内动态采光色度学质量研究

Chromatographic Quality Study of Dynamic Indoor Lighting with Spectrally Selective Windows

李森渊，薛鹏，沈意，张严匀，刘雨轩
Li Senyuan, Xue Peng, Shen Yi, Zhang Yanyun, Liu Yuxuan

摘要 光谱选择性窗户因其具备动态调控太阳辐射透射率的特性，已成为营造良好室内光环境和实现建筑节能的有效技术之一。然而，玻璃对太阳光谱的不均匀透射与吸收会改变透射日光的颜色质量，影响居住者的视觉和非视觉舒适度。为确保室内舒适的发光环境，全面评价光谱选择性窗户的动态采光色度学质量显得尤为重要。本研究基于日间动态辐射光谱，选用显色指数（CRI）和相关色温（CCT）两个指标综合评估了三种不同热致变色窗（TC）在北京地区应用的动态采光性能。结果表明，同时考虑到 CRI 和 CCT，TC 窗相较普通的双层玻璃窗年动态透射日光质量可提高近 7%。

关键词 光谱选择性窗户；动态辐射光谱；色度学；CCT；CRI

Abstract Due to its ability to dynamically adjust the solar radiation transmittance, spectrally selective windows have become one of the effective technologies for creating a good indoor light environment and realizing building energy saving. However, the uneven transmission and absorption of the solar spectrum by glass can alter the color quality of transmitted daylight, affecting the visual and non-visual comfort of occupants. A comprehensive evaluation of the dynamic chromatic quality of lighting from spectrally selective windows is particularly important in order to ensure a comfortable luminous environment indoors. In this study, the dynamic daylighting performance of three different thermochromic windows (TC) in Beijing was evaluated based on the dynamic daytime radiation spectrum, using two indicators: color rendering index (CRI) and correlated color temperature (CCT). The results show that considering both CRI and CCT, TC windows can improve the annual dynamic daylight quality by nearly 7% compared to ordinary double-glazed windows.

Keywords Spectrally selective windows; Dynamic radiation spectrum; Colorimetry; CCT; CRI

李森渊，北京工业大学，硕士研究生。
薛鹏，北京工业大学，校聘教授，博士生导师。
沈意，北京工业大学，硕士研究生。
张严匀，北京工业大学，博士研究生。
刘雨轩，北京工业大学，本科生。

“双碳”背景下城市夜景照明问题与对策探析

Analysis of Current Problems and Countermeasures for Urban Nightscape Lighting in the Context of Carbon Peaking and Carbon Neutrality

吕英慧，严永红
Lü Yinghui, Yan Yonghong

吕英慧，重庆大学建筑城规学院，硕士研究生。
严永红，重庆大学建筑城规学院，博士生导师，教授。

摘要 在气候变化、环境污染等问题成为时代热点的今天，一场席卷全球的保卫战正在同步进行。为了打赢这场保卫战，全球各国正致力于实现碳达峰和碳中和的新目标。在新形势下，城市夜景照明的发展面临着新挑战。城市夜景照明作为城市发展的产物，与气候环境问题有着千丝万缕的关系。城市夜景照明的良好发展对实现“双碳”的目标具有重要意义。本文从“双碳”的角度，深入剖析城市夜景照明现存的问题，并对当代城市夜景照明优化策略进行总结，由此为实现城市“双碳”目标提供借鉴。

关键词 “双碳”；夜景照明；优化策略；照明问题

Abstract At a time when climate change, environmental pollution and other issues have become the focus of the times, a global defence war is being waged. In order to win this defence war, countries around the world are committed to achieving the new targets of carbon peaking and carbon neutrality. Under the new situation, the development of urban night lighting is facing new challenges. As a product of urban development, urban nightscape lighting is inextricably linked to climate and environmental issues. The good development of urban nightscape lighting is of great significance in achieving the goals of carbon peaking and carbon neutrality. This paper analyses the existing problems of urban nightscape lighting in depth and summarises the optimisation strategies of contemporary urban nightscape lighting, so as to provide a reference for achieving the goals of carbon peaking and carbon neutrality in cities.

Keywords Carbon peaking and carbon neutrality; Night lighting; Optimization strategy; Lighting problem

---

地下轨道交通站点复合空间的照明引导性要素研究

Research on the Guiding Factors of Lighting in Composite space of the Underground Railway Transit Station

王立雄，黄博超，于娟，王楚尧
Wang Lixiong, Huang Bochao, Yu Juan, Wang Chuyao

王立雄，天津大学，研究员，博士生导师。
黄博超，天津大学，博士研究生。
于　娟，天津大学，助理研究员，教师。
王楚尧，天津大学，硕士研究生。

摘要 建筑照明内在的引导特性在塑造、规范空间上具有不可替代的作用。为提升复合空间的照明引导性，本文以地下轨道交通站厅空间为例，实地调研获取 29 个站厅空间的照明数据，采用 DIALux 软件开展照明仿真模拟，分析模拟数据、方案投票数和任务完成评分。结果表明，光源的空间布局要素与形态要素是影响照明引导性的关键要素。在此基础上，本文从分区照明及交错布置的光源空间布局要素、沿行走方向延伸的线型光源形态要素等方面提出照明优化策略，以期为建筑复合空间的照明引导性设计提供参考。

关键词 复合空间；照明引导性；影响要素；软件模拟

Abstract The inherent guiding characteristics of architectural lighting play an irreplaceable role in shaping and standardising space. In order to improve the lighting guidance of composite spaces, taking the hall space of underground railway transit station as an example, the lighting data of 29 station halls were obtained by field survey. Lighting simulation was carried out using DIALux software, and the simulation data, scheme votes and task completion scores were analysed. The results show that the spatial layout and the shape of the light source are the key factors affecting the lighting guidance. On this basis, the lighting optimisation strategy is proposed from the aspects of zoning lighting, staggered light source spatial layout elements, linear light source shape elements extending along the walking direction, etc., to provide a reference for the lighting guidance design of building composite space.

Keywords Composite space; Lighting guidance; Influencing factors; Software simulation

步行尺度下的商业街店面标识光环境亮度调查研究——以上海市南京路步行街为例

A Study of the Lighting Environment of Commercial Sign at the Pedestrian Scale—A Case Study of Nanjing East Road Commercial Street in Shanghai

曾宪宪，林怡，林天宇，
Zeng Xianxian, Lin Yi, Lin Tianyu

曾宪宪，同济大学建筑与城市规划学院，博士研究生。
林怡，同济大学建筑与城市规划学院，博士生导师。
林天宇，同济大学国豪书院，本科生。

摘要 过亮的商业街店面标识是夜间视觉污染之一，对人体的视觉、生理和心理产生负面影响，并导致不必要的电力消耗，不符合“双碳”的要求。以上海南京东路为例，本研究使用成像亮度计调查了商业街店面标识及周围环境的光环境亮度情况。结果显示，调查区域的标识亮度相对均匀，照明品质较高，但78%的标识超过了《城市夜景照明规范》中的最高亮度限值；店面标识的平均亮度与背景及周围环境相比差异较大，高亮度对比度将使行人产生严重的视觉不适感。本研究为未来城市夜景照明设计中的标识亮度控制提供建议，并强调了对商业街标识进行亮度管控的紧迫性。

关键词 商业标识；背景亮度；过度照明；光环境

Abstract Excessive brightness of commercial storefront signs contributes to nighttime visual pollution, which negatively affects human visual, physiological, and psychological well-being, and also leads to excessive electricity consumption, which contradicts the “carbon peaking and carbon neutrality” requirements. Taking East Nanjing Road in Shanghai as a case study, this research investgates the lighting conditions of commercial storefront signs and their surroundings using an imaging luminance meter. The results show that 78% of the surveyed sighs have a luminance exceeding the maximum limit. Although the signs have a relatively uniform luminance, there exists a significant disparity in the luminance ratio between the average sign luminance and that of the backgrounds and surroundings. This high luminance contrast can cause visual discomfort to pedestrians. The study provides recommendations for sign luminance control in future design and emphasizes the urgency of luminance regulation of commercial street signs.

Keywords Commercial sign; Background luminance; Excessive lighting; Lighting environment

道路监控补光灯对夜间行车安全影响研究

Research on the Influence of Monitoring Supplementary Light on Night Driving Safety

张昕烁，何荥，伍长生
Zhang Xinshuo, He Ying, Wu Changsheng

张昕烁，重庆大学建筑城规学院，博士研究生。
何荥，重庆大学建筑城规学院，教授。
伍长生，中南建筑设计院股份有限公司。

摘要 监控补光灯作为道路监控系统的一部分被广泛用于城市道路中，但监控补光灯由于其高亮度的特征，夜间容易在机动车驾驶员视野内形成眩光，影响行车安全。目前关于监控补光灯对行车安全的影响研究尚不充分，未能提出有效控制眩光的建议和策略。本文针对道路监控补光灯对夜间行车安全的影响进行了实验研究，为监控补光灯的眩光控制提出了控制策略。

关键词 监控补光灯；眩光；视觉表现；目标探测；反应时间

Abstract As a part of the road monitoring system, the monitoring supplementary light is widely used in urban roads. Because of its high brightness characteristics, it is easy to form glare in the field of vision of motor vehicle drivers at night, affecting driving safety. At present, the research on the impact of monitoring supplementary light on driving safety is not sufficient, and no suggestions and strategies for effective control of glare have been put forward. In this paper, the influence of monitoring supplementary light on night driving safety was studied experimentally, and the control strategy for glare control of monitoring supplementary light was proposed.

Keywords Monitoring supplementary light; Glare; Visual performance; Target detection; Reaction time

基于复合开窗系统的办公空间室内光环境优化研究

Optimization of the Daylight Environment in Office Space Based on Complex Fenestration Systems of Buildings

郭旭冉，赵永青，田真
Guo Xuran, Zhao Yongqing, Tian Zhen

郭旭冉，湖南大学建筑与规划学院，博士研究生。
赵永青，湖南大学建筑与规划学院，博士研究生。
田真，湖南大学建筑与规划学院，教授，博士生导师。

摘要 优化室内光环境能够节约建筑照明能耗且利于人体生理健康。本文基于模拟，分析了使用日光重定向棱镜膜和透光百叶的组合式复合开窗系统对提升室内光环境质量的潜力。研究表明：①本研究的复合开窗系统能够在保证室外视野的前提下，提高办公空间室内的采光均匀度，显著减少办公空间室内的眩光情况；②棱镜膜和 30% 穿孔百叶的组合形式对室内光环境的优化效果最好，$UDI_{300\sim3000}$、$GA_{40\%}$、OEV 都在 70% 以上。通过研究，本文探讨了复合开窗系统对营造办公空间健康光环境的可行性。

关键词 复合开窗系统；日光重定向；透光百叶；室内采光环境；模拟分析

Abstract Optimizing the indoor lighting environment can reduce energy consumption and promote human well-being. This study analyzed the potential benefits of a combined complex fenestration system, that integrates prismatic daylight redirecting fenestration and perforated louvers. The study revealed the following key findings: ① The implementation of this complex fenestration system can improve the uniformity of interior daylight in office spaces, effectively mitigating glare problems while preserving outdoor views; ② The combination of prismatic film and 30% perforated louvers has the best effect on optimizing the indoor daylighting environment, achieving $UDI_{300\text{-}3000}$, $GA_{40\%}$, and OEV over 70%. By exploring the feasibility of a complex fenestration system, this study provided insights into creating a healthy lighting environmenthin office spaces.

Keywords Complex fenestration systems; Daylight redirection; Perforated louvers; Daylighting luminance environment; Simulation and analysis

---

间接光眼部照度作为空间视亮度量化指标的适用性研究

Research on the Applicability of Indirect Corneal Illuminance as a Quantitative Parameter to Evaluate Spatial Brightness in Interior Spaces

胡治国，戴奇
Hu Zhiguo, Dai Qi

胡治国，复旦大学工程与应用技术研究院，博士研究生。
戴奇，复旦大学信息科学与工程学院光源与照明工程系，研究员，博士生导师。

摘要 空间视亮度与视觉舒适性密切相关，是影响建筑室内空间光品质的重要因素。目前，由于缺乏一个与主观评价高度相关又便于实际测量的指标，空间视亮度的量化十分困难。本文探讨了将间接光眼部照度作为空间视亮度量化指标的适用性：在具有不同光分布的间接光照明场景中，分别开展了一系列主观评价人因实验，获得了达到相同空间视亮度时各场景所对应的眼部照度数值。结果表明，即使光分布差异巨大，人们在各场景下达到相同空间视亮度时所需的间接光眼部照度十分接近。与 MRSE 和 $L_{av,B40}$ 指标相比，间接光眼部照度指标与被试的主观判断具有更好的相关性且测量更加便利。

关键词 空间视亮度；间接光眼部照度；光的空间分布；室内光环境；人因实验

Abstract Spatial brightness is related to the visual comfort and is an essential factor that affects the indoor lighting quality. Quantitative evaluation of spatial brightness has been difficult, mainly due to the lack of a metric that is both highly related to subjective evaluation and convenient to measure in the field. This work investigated the applicability of using indirect corneal illuminance to evaluate spatial brightness for a visual field in indoor environments. A series of human subjective evaluation experiments were carried out in various indirect-lighting scenes with different lighting distribution patterns. The corresponding values of indirect corneal illuminance required to achieve the same level of spatial brightness were obtained. The results showed that these corresponding values were very close, even under very different patterns of light distribution. Furthermore, compared MRSE and $L_{av,B40}$, it also had the best correlation with subjective evaluation of spatial brightness with the advantages of accuracy and convenience of measurement.

Keywords Spatial brightness; Indirect corneal illuminance; Spatial distribution of light; Interior luminous environment; Human factor study

人口老龄化背景下公路隧道照明节能策略

Strategy for Optimizing the Lighting Environment in Highway Tunnels Under the Background of the Aging Population

蔡贤云，田心怡，翁季
Cai Xianyun, Tian Xinyi, Weng Ji

摘要 人口老龄化背景下中老年驾驶人数量逐步增加，对现有公路隧道光环境要求更高。本文借助阈限亮度对比实验探讨了驾驶员年龄对隧道照明小目标可见度的影响作用，并提出人口老龄化背景下的照明节能策略。采用定值刺激法通过照明仿真实验，经过统计分析和修正系数得到不同年龄实验样本的阈限亮度差；并建立不同年龄组实验样本的阈限亮度差与小目标可见度水平之间的数学模型。结果表明：相同隧道光环境下青年组实验样本的小目标可见度高于中老年组。隧道照明设计必须考虑驾驶员的年龄因素。

关键词 隧道照明；节能策略；人口老龄化；小目标可见度；阈限亮度差

Abstract In an aging society, the number of middle-aged and elderly drivers is increased. Therefore, tunnel lighting environment is required to be higher for safe driving. In order to discuss the influence of driver' s age on small target visibility and the lighting environment optimization strategy in highway tunnel, luminance contrast threshold experiment was conducted. Based on the fixed value stimulation method, the minimum luminance threshold could be carried out by lighting simulation experiment. The minimum luminance threshold could be easily obtained by statistical analysis and correction coefficients. Therefore, the functional relationship between the minimum luminance threshold and the small target visibility was established under different age groups. The results show that the small target visibility of the young group is higher than that of the middle-aged and elderly groups under the same tunnel lighting environment, which indicatesthe age of driver must be considered in the tunnel lighting design.

Keywords Tunnel lighting; Lighting environment optimization strategy; Population aging; Small target visibility; Minimum luminance threshold

蔡贤云，重庆交通大学建筑与城市规划学院，副教授。
田心怡，重庆交通大学建筑与城市规划学院，硕士研究生。
翁季，重庆大学建筑城规学院，教授，副院长。

建筑天然光全年动态模拟优化方法

The Optimized Method for Annual Dynamic Daylight Simulation

赵永青，田真 *，郭旭冉
Zhao Yongqing, Tian Zhen, Guo Xuran

摘要 本文提出一种新的直射太阳光贡献计算方法，用之替换标准五步法中的太阳系数法，得到一种优化的五步法。通过将五步法和优化五步法与光线追踪法比较对该方法的结果进行验证。结果表明：与五步法（MF:3）相比，直射部分计算时间可以降低 71.9%，全过程计算时间可降低 34.5%。与五步法（MF:6）相比，直射部分计算时间可以降低 94.4%，全过程计算时间可降低 77.9%。在计算精度方面，即使是与五步法（MF:6）比较，优化五步法的眩光指标计算误差仍然小于五步法计算结果。

关键词 动态采光模拟；Radiance；眩光评估；优化五步法

Abstract This study proposes a new method for calculating direct solar contribution, which replaces the sun coefficient method in the standard five-step method and results in an optimized five-step method. The method was validated by comparing the five-step method and the optimized five-step method with ray-tracing method. The result shows that compared with the five-step method (MF:3), the calculation time of the direct solar part in the optimized five-step method can be reduced by 71.9%, and the whole process calculation time can be reduced by 34.5%. Compared with the five-step method (MF:6), the calculation time of the direct solar part can be reduced by 94.4%, and the whole process calculation time can be reduced by 77.9%. In terms of calculation accuracy, even when compared with the five-step method (MF:6), the calculated error of the glare index in the optimized five-step method is smaller than that in the five-step method.

Keywords Dynamic daylight simulation; Radiance; Glare evaluation; Modified five-phase method

赵永青，湖南大学建筑与规划学院，博士研究生。
田真，湖南大学建筑与规划学院，教授，博士生导师。
郭旭冉，湖南大学建筑与规划学院，博士研究生。

公路隧道光环境下驾驶员年龄差异研究

Study on Age Difference of Drivers Under Light Environment in Highway Tunnel

田心怡，蔡贤云 *，王晓钰，张悦，谭琳

Tian Xinyi，Cai Xianyun，Wang Xiaoyu，Zhang Yue，Tan Lin

田心怡，重庆交通大学建筑与城市规划学院，硕士研究生。
蔡贤云，重庆交通大学建筑与城市规划学院，副教授。
王晓钰，重庆交通大学建筑与城市规划学院，硕士研究生。
张悦，重庆交通大学建筑与城市规划学院，硕士研究生。
谭琳，重庆交通大学建筑与城市规划学院，硕士研究生。

摘要 中老年驾驶员的视觉特征与其余年龄段驾驶人群有明显不同，但在现有公路隧道照明研究中，缺乏对不同年龄驾驶员的综合考虑。本文针对公路隧道照明对各年龄段驾驶员的影响情况进行了问卷调研和现场实测工作，实验结果显示，中老年驾驶员的瞳孔直径在通过隧道入口段和结束段时较青年驾驶员有明显差异，结合调查问卷可以发现相较青年驾驶员，中老年驾驶员更易因为隧道光环境差异受到影响。文章针对这些问题，提出了相关改进意见，为今后公路隧道照明光环境设计和相关研究提供一些参考。

关键词 隧道照明；光环境；中老年驾驶员；视觉特征；瞳孔直径

Abstract The visual characteristics of middle-aged and elderly drivers are significantly different from those of other age groups, but in the existing research on highway tunnel lighting, there is a lack of comprehensive consideration of drivers of different ages. In this paper, the impact of highway tunnel lighting on drivers of all ages has been investigated by questionnaire and field measurement. The experimental results show that the pupil diameter of middle-aged and elderly drivers is significantly different from that of young drivers when passing through the entrance and end of the tunnel. Combining with the questionnaire, we can find that compared with young drivers, middle-aged and elderly drivers are more likely to be affected by the differences in tunnel lighting environment. In view of these problems, this paper puts forward relevant improvement suggestions, providing some reference for the future design and related research of lighting environment in highway tunnels.

Keywords Tunnel lighting; Light environment; Middle-aged and elderly drivers; Visual characteristics; Pupillary diameter

---

基于动态分析的大空间交通建筑天然光环境设计研究

Research on Daylighting Environment Design of Large Space Transportation Buildings Based on Dynamic Analysis

张滨，张恭铭，荆璐，张晟齐

Zhang Bin, Zhang Gongming, Jing Lu, Zhang Shengqi

张滨，建科环能科技有限公司，高级工程师。
张恭铭，建科环能科技有限公司，助理工程师。
荆璐，建科环能科技有限公司，助理工程师。
张晟齐，中国建筑科学研究院有限公司，硕士研究生。

摘要 航站楼、高铁站等交通建筑在我国快速发展，在该类建筑中合理利用天然光不仅可以提升室内光环境和旅客舒适度，同时也是提高建筑品质和实现低碳节能的重要途径。本文以北方某机场航站楼设计为例，研究交通建筑高大空间天然光导向下的天窗优化设计。运用光环境模拟软件，基于采光系数、动态光环境评价指标 UDI 以及晴天室内照度分布对天窗设计进行优化。通过对航站楼天然光环境的优化实践，为设计初期合理的天窗设计及室内天然光环境模拟提供参考依据。

关键词 大空间交通建筑；动态采光分析；天窗采光；采光模拟；优化设计

Abstract Transportation buildings such as terminals and high-speed railway station achieve rapid development in our country. Rational use of natural light can not only improve indoor light environment and passenger comfort, but also is an important way to improve building quality and low carbon energy saving. Taking the design of an airport terminal in the north as an example, this paper studies the optimization design of skylight guided down by natural light in main space. The skylight design was optimized by using the light environment simulation software based on the DF, UDI, and the indoor illumination distribution in sunny days. The optimization practice of natural light environment of terminal provides reference for reasonable skylight design and indoor natural light environment simulation at the initial stage of design.

Keywords Transportation building; Dynamic daylighting analysis; Skylight lighting; Daylighting simulation; Optimal Designing

光线入射角对圆筒式镜面导光管光传输效率的双重影响

Dual Influences of Light Incident Angle on Light Transmission Efficiencies of Cylindrical Mirror Light Pipes

欧阳金龙，王春苑，方舟，高庆龙
Ouyang Jinlong, Wang Chunyuan, Fang Zhou, Gao Qinglong

摘要 进入镜面导光管内的每束光线入射角不仅影响着其在管内反射次数，更决定了管内壁反射率。为了分析其对导光管光传输效率（LTEs）的双重影响，首先，基于光线追踪原理、镜面反射定律与空间立体几何，根据各光线进入管内的三种情况，建立了单束光线入射角与反射次数的计算式；其次，根据光的偏振理论，确定光线以不同入射角照射到金属镜面反射面会产生不同反射率；再次，探讨了光线入射角对导光管 LTE 的双重影响；最后，修正了任意光源下导光管 LTE 的表达式，为构建复杂光源下 LTE 数学模型，奠定了理论基础。

关键词 镜面反射；反射次数；反射率；光线追踪；光的偏振

Abstract The incident angle of every beam entering a specular light pipe not only affects its reflection times, but also determines the reflectivity. To analyze the dual influences of incident angle in the calculations of light transmission efficiencies (LTEs), firstly based on ray-tracing principle, specular reflection law and solid geometry, the expressions between incident angle and reflection time are established according to three cases of a single beam entering a pipe; secondly, according to light polarization theory, it is made clear that different reflectivity will be produced if the light strikes a metal specular surface at different incident angles; Thirdly, the dual influences of incident angle on any pipe's LTE is explored; finally, the LTE expression of a light pipe under any light source is modified for a theoretical foundation to establishing LTE mathematical models under complex light sources.

Keywords Specular reflection; Reflection time; Reflectivity; Ray-tracing; Light polarization

欧阳金龙，四川大学，教授。
王春苑，四川大学，实验师。
方舟，四川大学，实验师。
高庆龙，四川大学，教授。

非均匀穿孔遮阳板对室内光环境的影响——以武汉地区为例

Effects of Non-uniform Perforated Solar Screen on Indoor Lighting Environment—A Case Study of Wuhan Area

黄凌江，张小虎
Huang Lingjiang, Zhang Xiaohu

摘要 穿孔遮阳板作为一种常见的遮阳装置，被广泛用于现代建筑外立面。现有研究主要关注均匀穿孔遮阳板，而对非均匀穿孔遮阳板的研究有限。因此本文通过分析针对孔间距变化，进行特定趋势的随机分布、孔径尺寸变化，进行特定趋势的正交分布、穿孔率均匀递增变化，进行随机分布，三种不均匀变化的情况，探讨非均匀孔洞分布对室内光环境的影响。结果表明，不同非均匀孔洞分布形式下的穿孔遮阳板对实验模型光环境具有明显的影响。通过优化其疏密分布形态，能够减少眩光的产生并提高室内光环境。

关键词 非均匀穿孔遮阳板；眩光；室内光环境

Abstract The perforated solar screen, a commonly employed shading device, finds particular utility in solar shading, owing to its unique combination of baffle-type shading and light transmission properties. Existing studies predominantly concentrate on uniformly perforated solar screens, while investigations pertaining to non-uniformly perforated solar screen remain scarce. This paper aims to elucidate the impact of non-uniform hole distribution on the indoor lighting environment, achieved through an analysis of specific trends of random distribution, changes in aperture size, orthogonal trend distribution, uniform increases in perforation rate, and random distribution. The results show that perforated solar screens with varying non-uniform hole distributions markedly affect the light environment of the experimental model.

Keywords Non-uniform perforated solar screen; Glare; Indoor light environment

黄凌江，武汉大学城市设计学院建筑系，教授。
张小虎，武汉大学城市设计学院建筑系，硕士研究生。

基于WELL健康标准的居家办公智慧光环境VDT作业实验场景搭建与研究

Experimental Scenario Construction and Research on VDT Operation of Smart Luminous Environment Based on WELL Health Standard

姜书謦，史洁 *，赵颖珊，闫馨文，冯宇镝，胡惠杉
Jiang Shuxin, Shi Jie, Zhao Yingshan, Yan Xinwen, Feng Yudi, Hu Huishan

姜书謦，赵颖珊，闫馨文，冯宇镝，同济大学中德工程学院，本科生。
史洁，同济大学中德工程学院，副教授，导师。
胡惠杉，同济大学中德工程学院，实验室老师。

摘要 本项目基于 WELL 健康标准，以中医体质为分类标准，对疫情期间居家办公的上海居民进行了问卷调查，发现阴虚质和湿热质两种中医体质人群，较其余体质人群（平和质、气虚质等七种体质）受光照环境影响更为显著。另外，在实验室针对阴虚质和湿热质两类体质设计并搭建出了能够有效缓解其长时间 VDT 作业视疲劳的人工智慧光照场景，结果显示相比单一光照参数的光环境，实验组智慧光照环境可以缓解约 22.2% 的视疲劳症状加剧、减少约 32.4% 的视力下降出现频率。

关键词 VDT 作业；视疲劳；中医体质；光照场景

Abstract Based on the WELL Health Standard, this project conducted a questionnaire survey on Shanghai residents who worked at home during the epidemic, and found that two TCM constitutions, Yin Deficiency and Damp Heat, were more significantly affected by the light environment than the other seven constitutions (Ping He, Qi Deficiency, etc.). The results showed that compared to a single light environment, the experimental group was able to alleviate about 22.2% of the increase in visual fatigue symptoms and 32.4% of the decrease in the frequency of vision loss.

Keywords VDT work; Visual fatigue; TCM constitution; Luminous scenario

---

基于视觉与非视觉效应的商业中心区光环境评价研究——以南京新街口为例

An Investigation of Light Pollution Within Commercial Areas from Both Visual and Non-visual Perspectives —Take Xin Jie Kou in Nanjing as an example

孔哲，高一茗，王颖旭，李骏安，潘佳怡，张一欣，向畅颖，司秉卉
Kong Zhe, Gao Yiming, Zhang Yixin, Pan Jiayi, Wang Yingxu, Li Jun'an, Xiang Changying, Si Binghui

孔哲，东南大学，讲师。
高一茗、王颖旭、李骏安、潘佳怡、张一欣，东南大学，本科生。
向畅颖，香港科技大学，讲师。
司秉卉，南京工业大学，讲师。

摘要 商业区中心区的光环境普遍存在照明过亮、色彩繁杂、动态照明变化过快等问题。大量研究从光污染的视觉层面进行了探索，但光污染的非视觉生物效应层面有待探索。本文以南京新街口为例，实测了 6 个地点的夜晚光环境数据，将明视觉照度亮度转化为生物节律照度和亮度。同时采用问卷调研附近居民对光环境亮度、色彩等维度的评价。初步研究收集 52 份有效问卷。数据分析表明，6 个地点存在不同程度视觉和非视觉的光污染现象。

关键词 光污染；非视觉生物效应；视觉效应；高动态图片技术；问卷调查

Abstract Lighting environments within commercial areas cause pollutions due to over brightness, colorful sources and conversion speed. Many studies have explored lighting pollution within commercial areas from the visual perspective. However, lighting pollution on the non-visual effect requires further exploration. This field study uses Xin Jie Kou in Nanjing as an example. We measured both luminance and illuminance values at six representative spots. The data were later converted to circadian values. We also distributed a questionnaire to consumers to collect their evaluations on brightness, colorfulness and speed of lighting sources. We collected 52 valid questionnaires in this preliminary study. The results show that all six spots have severe lighting pollution issues from both visual and non-visual perspectives.

Keywords Lighting pollution; Non-visual effects; Visual effects; HDR-Technique; Survey

基于健康照明的住宅社区室外夜间光环境研究

Research on Outdoor Nighttime Light Environment of Residential Communities Based on Healthy Lighting

敖茂易，许子娴
AO Maoyi, Xu Zixian

敖茂易，重庆大学建筑城规学院，硕士研究生。
许子娴，苏州绕城高速公路有限公司，助理建筑师。

摘要 随着中国城市化率的提高和乡村建设的推进，住宅社区成为居民活动的主要区域，人们也越来越注重健康并追求健康的生活环境。光对人的心理和生理会产生巨大作用，住宅社区夜间光环境的改善迫在眉睫。本文通过实地测量社区的照明现状，调查住区居民对室外人工光环境的主观评价，分析得出室外光环境照明存在亮度不足和设备维护的问题。结合层次分析法，提出可通过夜间眩光控制、舒适照度、优化照明布局、照明光色调节等四个方面营造健康的室外夜间光环境。

关键词 健康照明；室外光环境；层次分析法；健康社区

Abstract With the increasing urbanization rate and rural construction in China, residential communities have become the main areas for residents' activities, and people are becoming more and more health conscious and pursuing a healthy living environment. Light has a huge effect on human psychology and physiology, and it is urgent to improve the nighttime light environment in residential communities. In this paper, we measured the lighting status of the community and investigated the subjective evaluation of outdoor artificial light environment by the residents, and analyzed the problems of insufficient brightness and equipment maintenance of outdoor light environment lighting. Combined with the AHP hierarchical analysis, it is proposed that a healthy outdoor nighttime light environment can be created through four aspects: night glare control, comfortable illumination, optimized lighting layout, lighting color adjustment.

Keywords Healthy lighting; Outdoor light environment; AHP; Healthy community

---

生态保护视角下人工光光谱对野外昆虫的趋光性影响规律

The Effect Patterns of Light Spectrum on Phototaxis of Insects in the Field from the Perspective of Ecological Conservation

郝庆丽，刘刚，王立雄，于娟，郁泽君，陈晓超
Hao Qingli, Liu Gang, Wang Lixiong, Yu Juan, Yu Zejun, Chen Xiaochao

郝庆丽，郁泽君，天津大学建筑学院，博士研究生。
刘刚，天津大学建筑学院，教授。
王立雄，天津大学建筑学院，教授。
于娟，天津大学建筑学院，助理研究员。
陈晓超，天津大学建筑学院，硕士研究生。

摘要 快速增长的夜间人工光在全球范围内对昆虫形成致命的吸引力。LED的光谱智能控制为调控人工光对昆虫的致命吸引力带来机遇与挑战，但潜在的光谱风险并未得到很好的测量评估。本文开展昆虫趋光实验，以趋光数量与趋光节律，探讨光谱的生态风险。总体来说，窄谱光的波长与吸引力成反比例关系，但在不同昆虫目间存在差异。短波蓝光吸引力最高，属于生态风险光谱；其次为中波绿光；长波红光吸引力最弱。高色温广谱光的吸引力更高，但与低色温相比差异不显著。在夜间光污染生态风险最严重的阶段，20:00—21:30是人工光吸引速度最快的阶段。

关键词 夜间人工光；光谱；LED；昆虫；吸引力

Abstract Rapidly increasing artificial light at night creates a fatal attraction to insects. The intelligent spectrum control of LEDs presents various challenges for reducing the fatal attraction of insects to artificial light. However, the potential risks and variations in their magnitude have not been precisely assessed. In this paper, we conduct phototaxis experiments to explore the ecological risks of the light spectrum by studying the number of captures of phototactic insects at different times and under different types of light. In general, the narrow-spectrum wavelengths were inversely proportional to attraction with variations among insect orders. Short-wave blue lights are the most attractive, which belong to the ecological risk spectrum. Followed by medium-wave green lights, which belongs relatively eco-friendly spectrum. And long-wave red lights, which belong to the absolute eco-friendly spectrum, have the weakest attraction. The attraction was stronger to 4500 K LEDs than to 3000 K LEDs, but the difference between them was insignificant. During the period with the highest risk, artificial light attracts insects at the fastest rate from 20:00 to 21:30.

Keywords Artificial light at night; Spectrum; LED; Insect; Attraction

会议室场景多视觉任务及其交替的光环境研究

Optimising the Lit Environment of Meeting Room to Make Multiple Visual Tasks Inclusive

缪晓波，苏航，赵思怡，张焱，杨彪
Miao Xiaobo, Su Hang, Zhao Siyi, Zhang Yan, Yang Biao

摘要 随着会议室中场景的多样化，会议室光环境中的视觉任务类型也相应增多。单一场景的照明方式无法同时满足纸质印刷媒介、视觉显示终端、人际交流和其他视觉任务对光环境的需求。本研究针对投影类视觉显示终端和纸质材料及其交替这三种视觉任务对象，在实验室环境中设置了 5 种投影内容亮度比和 4 种桌面照度的 20 种组合，由 20 名志愿者完成了绩效测评和主观问卷。数据分析表明，投影类视觉显示终端观看和纸质材料阅读这两种单一视觉任务所对应的最优光环境，并不是这两种视觉任务交替时的最优光环境。如要进一步营造兼容多视觉任务的会议室光环境，需要对每个视觉任务对应的光环境进行精细化设计。

关键词 多视觉任务；情景照明；视觉显示终端；会议场景

Abstract As meeting scenarios diversify, the types of visual tasks in meeting rooms increase accordingly. A single illumination method cannot simultaneously meet the light requirements of paper-based print media, visual display terminals, interpersonal communication and other visual tasks. In this study, 20 combinations of desktop illuminance and projection wall brightness were set up in a laboratory environment for each of these visual tasks, and 20 volunteers completed performance measures and subjective questionnaires. The data analysis showed that the optimal plan of light environment for multiple visual tasks was not put the plan of light environment for single visual task together simply. However, starting from the refinement of spatial light distribution, it is possible to make the conference room light environment compatible with multiple visual tasks.

Keywords Multiple visual tasks; Project display terminal; Light distribution; Conference scenes

缪晓波，赵思怡，哈尔滨工业大学（深圳），硕士研究生。
苏航，张焱，哈尔滨工业大学（深圳），博士研究生。
杨彪，哈尔滨工业大学（深圳），副教授，博士生导师。

LED 广告屏光污染分析及应用

Analysis and Application of Light Pollution of LED Advertising Screen

李悦，王沁芳，许鸣，蒋桂庆
Li Yue, Wang Qinfang, Xu Ming, Jiang Guiqing

摘要 本文在国内外光污染相关标准规范的调研基础上，将 LED 广告屏视为光源对其相关光学特性进行研究，通过实验室检验得到典型 LED 广告屏元件的配光曲线及光通量，再结合 DIALux 软件进行模拟，研究 LED 广告屏在不同参数下对周边建筑的光侵扰程度。最终结合工程实例，及其光污染标准规范，模拟得到 LED 广告屏参数限值，并对安装应用后的 LED 广告屏使用情况进行跟踪检测，通过实际检测数据进行优化调整，降低高亮度 LED 广告屏对周边环境的光污染。

关键词 光污染；LED 广告屏；DIALux；模拟

Abstract Based on the investigation of relevant standards and specifications of light pollution at home and abroad, this paper studies the relevant optical characteristics of LED advertising screen as a light source, obtains the light distribution curve and luminous flux of typical LED advertising screen components through laboratory tests, and then simulates with DIAlux software to study the light intrusion degree of LED advertising screen on surrounding buildings under different parameters. Finally, combined with project examples and light pollution standard, the parameter limit of LED advertising screen is simulated, and the use of LED advertising screen after installation and application is tracked and tested, and by means of actual detection data, the parameters of LED advertising screen is optimized and adjusted to reduce the light pollution of the high brightness LED advertising screen to the surrounding environment.

Keywords Light pollution; LED advertising screen; DIALux; Simulation

李悦，常州市建筑科学研究院集团股份有限公司，助理工程师，绿色检测咨询组组长。
王沁芳，常州市建筑科学研究院集团股份有限公司，研究员级高级工程师，绿色节能中心副主任兼集团技术服务中心暖通专业总工。
许鸣，常州市建筑科学研究院集团股份有限公司，绿色节能中心主任，研究员级高级工程师。
蒋桂庆，常州市建筑科学研究院集团股份有限公司，检测员。

基于建筑模型探索室内环境采光的实验方法

Research on Experimental Method of the Indoor Daylighting Environment Under Model Building

郑华美，董书杨，陈玥杉，颜楚方，沈唐凛
Zheng Huamei, Dong Shuyang, Chen Yueshan, Yan Chufang, Shen Tanglin

郑华美，福州大学建筑与城乡规划学院，讲师，博士。
董书杨，凯达环球建筑设计咨询（北京）有限公司，助理建筑师，硕士研究生。
陈玥杉，华侨大学建筑学院，博士研究生。
颜楚方，集美区发展和改革局，科员。
沈唐凛，常州艺涧文化传媒有限公司，设计师。

摘要 为了充实建筑室内光环境的实验教学模式，倡导使用建筑模型探索室内光环境特性的教研平台。仿造适用于在人工天穹光气候条件下，多变量采光口结构的建筑模型，从中探究测量人工天穹亮度和建筑模型内照度的实验方法和原理。在建筑模型采光口上沿高度不变的情况下，测得不同窗台高的进深采光系数分布轨迹。在本建筑模型实验范围内，显示了使用建筑模型进行实验教研是可行的，从而为不同采光口和界面作用下的探究奠定了基础。

关键词 单侧窗；建筑模型；人工天穹；照度；采光系数

Abstract This paper is to explore the experimental method and characteristics of the indoor daylighting environment under the model building for course teaching. The model building has been imitated much variable quantity of structure of the windows for measuring daylight factor through using the equipment of artificial sky hemisphere. The measuring method and theory have been explored through measuring the equipment' s brightness and the indoor illuminance of the model building. While the upper edge height of windows is always the same, the daylight factor of depth direction has been measured for changing the window sill height. Under this model building, its result shows that the model building can be applicable for the experimental teaching. At the same time, the study will be expanded for changing the different of the interior surface or the window forms.

Keywords One side windows; Model building; Artificial sky hemisphere; Illuminance; Daylight factor

---

产品碳足迹分析相关标准对比及其照明应用思考

Comparison and Application Analysis of Standards for Carbon Footprint of Lighting Product

高雅春，罗涛，贾政焱
Gao Yachun, Luo Tao, Jia Zhengyan

高雅春，建科环能科技有限公司，工程师。
罗　涛，建科环能科技有限公司，研究员。
贾政焱，中国建筑科学研究院有限公司，助理工程师。

摘要 为应对气候变化，减少温室气体排放已经成为世界各国的重要共识。在这样的背景下，我国提出了争取 2030 年前碳达峰、2060 年前碳中和的庄严承诺。我国照明用能成为全社会总用电量的重要组成部分，进一步做好照明领域的节能减排也是确保“双碳”目标实现的重要途径之一。开展照明产品的碳足迹分析也将为温室气体减排起到重要的支撑作用。本文通过对现有产品碳足迹分析相关的标准进行对比分析，提出了照明产品碳足迹核算评估的应用实施的一些思考。

关键词 碳足迹；照明产品；标准化；全生命周期

Abstract The reduction of greenhouse gas emissions has become an important consensus among countries around the world to cope with the climate change. In this context, China has made a solemn commitment to reach the carbon peak by 2030 and carbon neutral by 2060. China' s lighting energy consumption has become an important part of the total electricity consumption of the whole society. Energy saving and GHG emission reduction in the field of lighting is also one of the important ways to ensure the realization of the “double carbon” goal. The carbon footprint analysis of lighting products will also play an important role in supporting the reduction of greenhouse gas emissions. This paper presents some thoughts on the application and implementation of the carbon footprint assessment of lighting products by comparing and analyzing the existing standards related to the carbon footprint analysis of products.

Keywords Carbon footprint; Lighting product; Standardization; Life cycle

第 4 篇
绿色建筑环境品质

高铁站候车厅热光声环境因子对舒适度的交互作用研究

Research on the Interaction of Thermal/Light/Acoustic Environmental Factors on Comfort in the Waiting Hall of High-Speed Railway Station

杜晓辉，张永超
Du Xiaohui,
Zhang Yongchao

杜晓辉，安徽建筑大学安徽省建筑声环境重点实验室，北京交通大学建筑与艺术学院，副教授。
张永超，北京市市政工程设计研究总院有限公司，工程师。

摘要 人们对室内环境的舒适度感受，是多种环境因子交互作用下的综合反映。本研究重点关注高铁客站候车厅中温度、照度、声压级的交互作用对人体环境舒适感知的影响。选取北方寒冷地区大型高铁客站，采用环境测试与问卷调查结合的方法，开展室内物理环境质量与旅客满意度的测试与调研，包括热环境、光环境、声环境等，对高铁客站室内环境质量与旅客满意度进行对比，并结合主客观测试结果展开综合分析。同时，将环境主观舒适评价与人体客观生理指标结合测试，对研究主题进行探讨。研究结论为高铁客站优化建筑空间设计与室内环境质量、提高乘客环境舒适度提供量化参考。

关键词 候车厅；环境舒适度；交互作用；室内环境因子

Abstract People' s feelings of comfort about the indoor environment are a comprehensive reflection of the interaction of various environmental factors. This paper focuses on the impact of the interaction of air temperature, illumination and sound pressure level on human environment comfort perception in the waiting hall of high-speed railway stations. In large high-speed railway stations in the cold area, with the method of environmental testing and questionnaire survey, the indoor environmental measurements and passengers' satisfaction survey were conducted, covering indoor thermal environment, lighting environment, acoustic environment and air quality. The paper analyses and compares indoor environmental quality(IEQ) and passengers' satisfaction among different railway stations. The research conclusion provides quantitative data for improving the passenger environmental comfort.

Keywords Waiting hall; Environment comfort; Interaction; Indoor environmental factors

---

寒冷地区绿色高层办公建筑自然通风系统性优化设计

Systematic Optimization Design of Natural Ventilation for a Green High-rise Office Building in the Cold Region

李琪，吉喆，倪树清，王慧，孟岩岩
Li Qi, Ji Zhe, Ni Shuqing,
Wang Hui, Meng Yanyan

李琪，同圆设计集团股份有限公司，中级工程师。
吉喆，同圆设计集团股份有限公司，高级工程师，绿建所所长。
倪树清，同圆设计集团股份有限公司，高级工程师。
王慧，同圆设计集团股份有限公司，中级工程师。
孟岩岩，同圆设计集团股份有限公司，助理工程师。

摘要 当前，高层办公建筑人员密度大、运行时间长，室内空气品质对建筑内人员工作效率具有较大影响，以新风系统为主的通风换气，造成较高空调能耗。本文以寒冷地区某绿色三星级高层办公建筑为研究对象，利用Phoenics软件进行模拟分析论证，顺应过渡季风向，从建筑体块，室内空间营造，平面布局进行系统性优化探索，形成引风通廊，优化通风路径，营造良好的自然通风环境。

关键词 绿色高层办公建筑；自然通风；模拟分析；建筑体块；空间营造；平面布局

Abstract At present, high-rise office buildings have high personnel density and long operation hours. Indoor air quality has a great impact on the working efficiency of people in building, that mostly depends on the fresh air system, resulting in high energy consumption of air conditioning. In this paper, a green three-star high-rise office building in the cold region is taken as the research object, the software Phoenics is used for simulation analysis and demonstration. Conforming to the transition monsoon direction, systematic optimization and exploration are carried out from the aspects of building block, indoor space construction, and plane layout, so as to form a ventilation corridor, optimize the ventilation path and create a good natural ventilation environment.

Keywords High-rise office buildings; Natural ventilation; Simulation analysis; Transition season; Systematic optimization

既有棚户区自然通风改造的策略与模拟研究
——以呼和浩特为例

Strategy and Simulation Study of Natural Ventilation Retrofitting in Existing Shantytowns —Take Hohhot as a Case

李果，陶志毅，王崴
Li Guo, Tao Zhiyi, Wang Wei

摘要 既有棚户区绿色改造中，如何改善住区环境，提高室内环境尤为重要。文章选取的研究对象为呼和浩特市前八里庄村的两个网格棚户区。首先梳理了我国棚户区改造的发展现状和应对策略，简述了自然通风的基本原理，并陈述了相关研究和本研究中选择自然通风的原因。随后，通过实地调查和数据收集，分析了影响住宅区内外空间通风质量的因素，提出基于自然通风技术下优化户型布局、改善其室内通风条件的改造策略与方案。最后借助软件模拟分析和比较通风条件的绿色策略和方案，验证改造后的改善效果。

关键词 自然通风；棚户区；既有改造；策略研究；软件模拟

Abstract It is especially important to improve the environment of settlements and the indoor environment in the green transformation of existing shantytowns. The research object selected in the article is two grid shantytowns in the former Bailizhuang village of Hohhot. Firstly, the development status and response strategies of shantytown renovation in China are sorted out, the basic principles of natural ventilation are briefly described, and the reasons for selecting natural ventilation in related studies and this study are stated. Then, through field survey and data collection, the factors affecting the ventilation quality of internal and external spaces in residential areas are analyzed, and the renovation strategies and solutions for optimizing the layout of household types and improving their indoor ventilation conditions based on natural ventilation technology are proposed. Finally, the green strategies and solutions for ventilation conditions are analyzed and compared with the help of software simulations to verify the improvement effects of the retrofit.

Keywords Natural ventilation; Shantytown; Existing renovation; Strategy study; Software simulation

——
李果，内蒙古工业大学建筑学院，硕士研究生。
陶志毅，内蒙古工业大学建筑学院，硕士研究生。
王崴，内蒙古工业大学建筑学院，副教授。

高大空间风口类型对室内环境的影响

Influence of the Supply Air Diffuser Choice on Indoor Environment in Tall Space

吉玉辰，郑艺倩，姜峰，朱雪涛，徐琴华，陈潇，沈鹏元 *
Ji Yuchen, Zheng Yiqian, Jiang Feng, Zhu Xuetao, Xu Qinhua, Chen Xiao, Shen Pengyuan

摘要 空调系统风口有多种类型，本文通过计算流体力学模拟方法比较分析了 3 种常用风口类型对室内环境的影响。研究发现，夏季工况下，在人行高度区域（z=1.5 m），对于 8 m 和 10 m 高的建筑，旋流风口的平均空气温度较舒适，比条形和球形风口降低约 2.5 ℃，同时旋流风口空气龄较小，综合效果最佳；冬季工况下，在人行高度区域，在 8 m 高空间中，旋流和条形风口室内平均温度较高（约 22.5 ℃），球形风口稍差（约 17.5 ℃），且条形和旋流风口空气龄较小，综合考虑上述结果，在高大空间中，优先推荐条形和旋流风口。

关键词 风口类型；室内热环境；热舒适；空气龄

Abstract There are various types of air diffusers utilized in air conditioning systems. In this study, we compare and analyze the influence of three different diffusers on the indoor environment using CFD simulation. The results revealed that during summer conditions, the swirl diffuser provided the most comfortable average temperature in the pedestrian area (z=1.5 m), with a 2.5 °C lower temperature than the strip and spherical diffusers in spaces that were 8-meter- and 10-meter- high. In addition, the swirl diffuser exhibited a low age of air, resulting in better overall performance. During winter conditions, the average temperature of the swirl and strip diffusers was higher (about 22.5 °C) in the pedestrian area, while the spherical diffuser was slightly lower (about 17.5 °C) in the 8-meter-high space. Both the strip and swirl diffusers demonstrated low ages of air. Based on these findings, we recommend selecting the strip and swirl diffusers as the preferred choice.

Keywords Air outlet type; Indoor thermal environment; Thermal comfort; Age of air

——
吉玉辰，哈尔滨工业大学（深圳）建筑学院，博士研究生。
郑艺倩，哈尔滨工业大学（深圳）建筑学院，硕士研究生。
姜峰，深圳市杰恩创意设计股份有限公司，董事长。
朱雪涛，徐琴华，深圳市杰恩创意设计股份有限公司，部门总监。
陈潇，深圳市杰恩创意设计股份有限公司，董事长助理。
沈鹏元，哈尔滨工业大学（深圳）建筑学院，副教授。

基于自适应热舒适的混合模式建筑自然通风运行区间

Natural Ventilation Operation Range for Mixed-mode Buildings Based on Adaptive Thermal Comfort

于瑛，相天慧，王迪，杨柳
Yu Ying, Xiang Tianhui, Wang Di, Yang Liu

于瑛，西安建筑科技大学，副教授，硕士生导师。
相天慧，西安建筑科技大学，硕士研究生。
王迪，西安建筑科技大学，硕士研究生。
杨柳，西安建筑科技大学，教授，博士生导师。

摘要　混合模式建筑结合自然通风和传统机械加热冷却系统，合理控制自然通风可降低建筑空调能耗并保证室内热舒适性。基于此，本文提出一种变阈值的自然通风运行区间的优化方法。首先，增设阈值条件对常用自然通风运行区间进行扩充，并依据自适应模型舒适极限将其划分为 9 个自然通风运行子区间。其次，通过联合仿真分析每个运行子区间的舒适性及能耗表现，以组合出适用于当地气候的最优自然通风运行区间。最后，以西安地区的混合模式建筑为仿真案例，结果表明根据本文所提优化方法制定的自然通风运行区间，可在不牺牲热舒适的前提下大幅降低能耗约 13%~45%。

关键词　混合模式；自然通风；热舒适性；节能潜力

Abstract　Mixed-mode buildings combine natural ventilation and traditional mechanical heating and cooling systems, and reasonable control of natural ventilation can reduce building air conditioning energy consumption and ensure indoor thermal comfort. Based on this, an optimization method of natural ventilation operation range with variable threshold is proposed in this paper. Firstly, the common natural ventilation operation range is expanded by adding threshold conditions and divided into nine natural ventilation operation subrange based on the comfort limit of the adaptive model. Secondly, the comfort and energy consumption performance of each operating subranges is analyzed by co-simulation to combine the optimal natural ventilation operating range for the local climate. Finally, using a mixed-mode building in Xi'an as a simulation case, the results show that the natural ventilation operation range developed according to the optimization method proposed in this paper can significantly reduce energy consumption by about 13%~45% without sacrificing thermal comfort.

Keywords　Mixed-mode; Natural ventilation; Thermal comfort; Energy-saving potential

---

自然通风调控下的北方高校既有宿舍楼环境优化研究——以内蒙古工业大学为例

Environmental Optimization of Existing Dormitory Buildings in Northern Universities Under Natural Ventilation Regulation—Take Inner Mongolia University of Technology as an Example

林裳，王婷 *
Lin Shang，Wang Ting

林裳，内蒙古工业大学建筑学院，硕士研究生。
王婷，内蒙古工业大学建筑学院，讲师。

摘要　高校既有学生宿舍楼改造是推进宿舍楼可持续性发展、提升学生居住环境、激活校园活力的重要途径，在既有学生宿舍楼改造中，生活环境优化尤为重要。文章首先梳理了北方高校既有宿舍楼普遍存在的问题，分析得出自然通风问题是其中的主要问题；其次选取内蒙古工业大学 2 栋学生宿舍楼为研究对象，通过实地调研及软件模拟的方法，分析影响宿舍楼公共空间自然通风的因素，并基于自然通风提出优化公共空间、改善通风条件的绿色策略；最后借助软件模拟验证改造后的提升效果。

关键词　自然通风；宿舍楼改造；策略研究；软件模拟

Abstract　The renovation of existing student dormitory buildings in colleges and universities is an important way to promote the sustainable development of dormitory buildings, improve students' living environment and activate campus vitality. In the renovation of existing student dormitory buildings, the optimization of their living environment is particularly important. Firstly, this paper combs the common problems of existing dormitory buildings in northern universities, and analyzes that natural ventilation is the main problem. Secondly, two student dormitory buildings of the Inner Mongolia University of Technology were selected as the research objects. Through field investigation and software simulation, the factors affecting the natural ventilation of the public space of the dormitory building were analyzed. Based on natural ventilation, a green strategy for optimizing public space and improving ventilation conditions was proposed. Finally, the software simulation is used to verify the improvement effect after the transformation.

Keywords　Natural ventilation; Dormitory building renovation; Strategy study; Software simulation

太阳能烟囱及热源对多层建筑室内热环境影响研究

Research on the Influence of Solar Chimneys and Heat Source on the Indoor Thermal Environment of Multi-storey Buildings

蔡阳，董明浩，黄晓燕，刘子瑞，赵福云
Cai Yang, Dong Minghao, Huang Xiaoyan, Liu Zirui, Zhao Fuyun

摘要 太阳能烟囱技术在强化自然通风和提高室内热舒适性等方面具有重要作用，是提高太阳能利用率的有效途径。本文结合太阳能烟囱的通风特性，探究太阳能烟囱及室内热源对多层建筑室内环境的影响。构建太阳能烟囱与室内热源耦合的通风模型，分析不同太阳能光照强度及内热源强度对多层建筑室内热环境的影响，并探究多层建筑自然通风量的变化规律。研究结果表明：随着太阳辐射强度的增大，太阳能烟囱通风量逐渐增大。本文将为太阳能烟囱强化自然通风技术在具有高温内热源建筑中的应用提供参考和依据。

关键词 太阳能烟囱；太阳辐射强度；自然通风；内热源强度

Abstract Solar chimney technology plays an important role in strengthening natural ventilation and improving indoor thermal comfort, and is an effective way to improve solar energy utilization. Based on the ventilation characteristics of solar chimneys, this paper explores the impact of solar chimneys and indoor heat source on the indoor environment of multi-storey buildings. Firstly, a ventilation model coupled with a solar chimney and indoor heat source is constructed. Secondly, the impact of different solar irradiance intensities and internal heat source intensities on the indoor thermal environment of multistory buildings is analyzed, and the variation law of natural ventilation capacity in multistory buildings is explored. The results show that with the increase in solar radiation intensity, the ventilation capacity of the solar chimney gradually increases. This study will provide a reference and basis for the application of solar chimney enhanced natural ventilation technology in buildings with high-temperature internal heat sources.

Keywords Solar chimney; Solar radiation intensity; Natural ventilation; Inner heat source intensity

蔡阳，暨南大学国际能源学院，副教授。
董明浩，刘子瑞，暨南大学国际能源学院，本科生。
黄晓燕，暨南大学国际能源学院，硕士研究生。
赵福云，武汉大学动力与机械学院，教授。

冬季教学建筑室内空气品质影响研究

Study on the Impact of Indoor Air Quality in Winter

蒋婧，聂皓清
Jiang Jing, Nie Haoqing

摘要 教室内空气品质与学生的身心健康和学习状态密切相关。在冬季供暖期选取西安某大学内 9 个典型教室，对室内环境参数进行监测，同时采用问卷调查对 375 名学生对室内空气品质进行主观评价。结果表明：在连上四节课的教室内，温、湿度均符合国家标准，CO_2 浓度平均每小时最高可达 2052 ppm，远超我国室内空气质量标准要求；学生对空气新鲜程度的变化十分敏感，人员密集高、空气流动频率小是使学生满意度与舒适度降低的主要原因；出现犯困、皮肤干燥和烦躁症状的人数变多且症状较严重，同时受试者的学习效率、记忆均下降。

关键词 室内空气品质；CO_2 浓度；SBS 症状；主观评价

Abstract Indoor air quality in classrooms is closely related to students' physical and mental health and learning status. Nine typical classrooms in a university in Xi'an were selected to monitor indoor environmental parameters during the winter heating period, while a questionnaire was used to subjectively evaluate the indoor air quality of 375 students. The results showed that in the classrooms where four classes were held in a row, the temperature and humidity met the national standards, while the average hourly maximum CO_2 concentration can reach 2052 ppm, far exceeding the requirements of China's indoor air quality standards; students were sensitive to changes in the freshness of the air, and the high density of people and low frequency of airflow were the main reasons for reducing student satisfaction and comfort; symptoms such as poor breathing, irritability, and sleepiness increased with the increase in study time The number of symptoms of drowsiness, dry skin, and irritability increases and the symptoms are more severe, while the subjects' learning efficiency and memory decreased.

Keywords Indoor air quality; CO_2 concentration; SBS symptoms; Subjective evaluation

蒋婧，西安工程大学，副教授，硕士研究生导师。
聂皓清，西安工程大学，硕士研究生。

适老病房物理环境调节措施效果实验研究

Effect of Adjustment Measures on the Physical Environment of Senile Wards

赵光颖，孟庆林，毛会军
Zhao Guangying, Meng Qinglin, Mao Huijun

摘要 本文以广州市某适老病房为研究对象，通过实地测试明确了吸顶风扇和外窗遮阳卷膜对室内物理环境的调控效果。结果表明，风扇不会使房间内温度有明显变化，适合在日间静坐或静卧时配合开窗或空气净化器使用。遮阳卷膜在中午时段降温效果最佳，北向房间空气温度最大可降低 0.27℃，宜在太阳辐射较大或老人日间需睡眠时使用。同一环境中，人体不同活动状态的热舒适差别较大，空调设定温度高低排序应为静坐状态 > 睡眠状态 > 躺卧状态。该研究可为适老病房环境调节设备选择及调控提供参考。

关键词 适老病房；物理环境调节；老年人；热舒适

Abstract Taking a senile ward in Guangzhou as the research object, this paper carried out indoor environmental quality testing experiments in summer, clarified the influence and regulation effect of the indoor environment of heat insulation film and fan, and evaluated the improvement effect of different environmental adjustment measures on human thermal comfort by using thermal comfort indicators. The results show that the fan does not cause significant temperature changes in the room, and is suitable for use with windows or air purifiers when sitting or lying still during the day. The shading film has the best cooling effect at noon, and the maximum temperature can be reduced by 0.27°C in the north room. It should be used during the day when the solar radiation is intense or the elderly need to sleep. In the same environment, the thermal comfort of the human body varies greatly in different activity states. The air conditioning temperature set for the sitting state, the sleeping state, and the lying state is lowered in order. These results can provide a reference for the selection and regulation of environmental regulation equipment in senile wards.

Keywords Senile ward; Environmental regulation; The elderly; Thermal comfort

赵光颖，华南理工大学建筑学院，硕士研究生。
孟庆林，华南理工大学建筑学院，教授，亚热带建筑科学国家重点实验室副主任，中国建筑学会建筑物理分会理事长。
毛会军，华南理工大学建筑学院，博士研究生。

石膏基调湿材料调湿性能试验研究

Experimental Study on Humidity Control Performance of Gypsum-Based Humidity-Control Material

李曦，冉茂宇
Li Xi, Ran Maoyu

摘要 在石膏基硅胶调湿复合材料的基础上，掺入质量比为 20% 的氯化钙改性海泡石粉，制备了新型石膏基调湿材料，并对其调湿性能进行了测试与分析。首先在封闭的小型试验箱内对材料进行除湿和放湿性能研究，而后分析了环境温度、相对湿度对其调湿性能的影响，并以等温吸放湿曲线作为评价指标，进一步研究了石膏基调湿材料的调湿性能。结果表明，该材料具有较好的调湿性能，且环境温湿度对材料调湿性能皆有影响，测得石膏基调湿材料的最大平衡含湿量为 0.225 g/g，是石膏基硅胶复合调湿材料的 1.4 倍，是纯石膏材料的 4.5 倍。

关键词 湿度控制；复合材料；石膏；硅胶；海泡石

Abstract This paper presents a novel type of gypsum-based humidity control material based on gypsum-silica gel humidity control material prepared by adding 20% sepiolite power activated by calcium chloride and its humidity control performance was conducted in experiments studies. The dehumidification and humidification characteristics of the material were studied in a test chamber. The impact of environmental temperature and relative humidity on moisture regulation was analyzed. Isothermal adsorption-desorption curves were used to evaluate the humidity control performance. Results show that the gypsum-based material absorbs moisture in high-humidity environments and releases moisture in low-humidity environments. Both temperature and humidity affect the material's performance. At 97.4% relative humidity, the material has a maximum equilibrium moisture content of 0.225 g/g, which is 1.4 times higher than the gypsum-silica gel material and 4.5 times higher than pure gypsum.

Keywords Humidity control; Composite material; Gypsum; Silica gel; Sepiolite power

李曦，华侨大学建筑学院，博士研究生。
冉茂宇，华侨大学建筑学院，教授。

热景观视角下既有教育建筑外遮阳设计研究

Research on External Sunshade Design of Existing Educational Buildings from the Perspective of Thermalscape

李琪，宋德萱
Li Qi, Song Dexuan

摘要 为获得既有教育建筑遮阳改造后的室内环境数据，以上海地区两栋教学楼为例，对教室空间进行光环境、风热环境实测，并利用 Grasshopper 平台进行参数化建模，计算各形式遮阳构件的最佳安装参数。结果表明，水平遮阳在室内光热环境方面的表现优于穿孔板；水平旋转式外遮阳具有较大的节能潜力与较好的适用性；穿孔板节能效果不显著。最后从热景观营造角度对既有教育建筑外遮阳改造提出建议。

关键词 热景观；教育建筑；外遮阳；建筑能耗

Abstract In order to obtain the indoor environmental data of the existing educational buildings after shading transformation, two teaching buildings in Shanghai were taken as examples, the light environment and wind and heat environment of the classroom space were measured, and the parametric modeling was carried out using the Grasshopper platform to calculate the best installation parameters of various forms of shading components. The results show that horizontal shading performs better than perforated panels in indoor light and thermal environments; The horizontal rotating external sunshade has great energy-saving potential and good applicability; The energy-saving effect of perforated panels is not significant. Finally, suggestions are made for the external shading renovation of existing educational buildings from the perspective of thermal landscape creation.

Keywords Thermalscape; Educational buildings; External shading; Building energy consumption

李琪，同济大学建筑与城市规划学院，博士研究生。
宋德萱，同济大学建筑与城市规划学院，教授，博士生导师，国家一级注册建筑师。

围合式教学楼朝向对室内风环境和 CO_2 浓度的影响研究

Research on the Influence of the Orientation of the Enclosed Teaching Building on the Indoor Wind Environment and the Distribution of CO_2

徐燊，王明豪
Xu Shen, Wang Minghao

摘要 教学楼室内人员密集，空气质量问题频发，充足的自然通风至关重要。本文旨在研究建筑朝向对围合式教学楼室内自然通风和 CO_2 浓度的影响。以武汉地区为例，开展了高校教学楼室内 CO_2 浓度的实测研究，利用 CFD 对不同朝向的围合式教学楼原型进行了自然通风和 CO_2 浓度的模拟，并进行了规律总结。研究发现，冬季高校教学楼室内普遍存在 CO_2 浓度超标现象，这和室内自然通风不畅有关；不同朝向下，围合式教学楼室内风环境和污染物分布变化规律不同。

关键词 高校教学楼；自然通风；围合式建筑；室内平均 CO_2 浓度；CFD

Abstract Sufficient natural ventilation is paramount in teaching buildings with high population densities and recurring indoor air quality issues. This study aims to examine the impact of building orientation on indoor natural ventilation and CO_2 concentration within enclosed teaching buildings. Aiming at the Wuhan area, this paper conducted a measurement study of the concentration of CO_2 in college teaching buildings and employed the CFD to simulate natural ventilation and CO_2 concentration in prototypes of enclosed teaching buildings with varying orientations, from which the fundamental laws were derived. The study revealed that during winter, the indoor concentration of CO_2 in college teaching buildings typically exceeds the recommended level due to personnel activities. The indoor wind environment and pollutant distribution patterns within enclosed teaching buildings differ under various orientations.

Keywords University teaching buildings; Natural ventilation; Enclosed buildings; Indoor average CO_2 concentration; CFD

徐燊，华中科技大学建筑与城市规划学院，湖北省城镇化工程技术研究中心，教授，博士生导师。
王明豪，华中科技大学建筑与城市规划学院，湖北省城镇化工程技术研究中心，博士研究生。

寒冷地区过渡季节教室内热舒适与空气品质调查研究

An Investigation on Thermal Comfort and Air Quality in Classroom During Transition Season in Cold Area

蒋婧，刘菲，云旭鑫
Jiang Jing, Liu Fei, Yun Xuxin

摘要 随着人们对学习效率的日益关注，学习空间的物理环境质量越来越受到重视。本文以寒冷地区某高校教室为研究对象，对过渡季教室内的热环境参数与环境质量参数进行测量，同时以调查问卷的形式对 340 名学生的穿衣情况以及他们对教室环境的热感觉评价进行研究。结果显示：西安市过渡季学生群体平均服装热阻为 0.77 clo，教室过渡季实测和预测热中性温度分别为 20.2℃和 18.4℃，90% 的学生感到满意的舒适温度区间为 20.1~25.3℃。教室内环境不太满足学生对热舒适和空气品质的需求，学生适应性行为有加减衣服、轻度活动、开门窗通风等。

关键词 寒冷地区；过渡季；热舒适；空气品质

Abstract As people pay more and more attention to learning efficiency, more and more attention is paid to the physical environment quality of learning space. Taking a university classroom in a cold region as the research object, this paper measured the thermal environment parameters and environmental quality parameters in the classroom during the transition season, and studied the dressing condition of 340 students and their thermal sensation evaluation of the classroom environment in the form of a questionnaire. The results show that the average thermal resistance of clothes of students in Xi'an during the transition season is 0.77 clo, the measured and predicted thermal neutral temperatures of classrooms during the transition season are 20.2°C and 18.4°C, respectively, and 90% of students are satisfied with the comfortable temperature range of 20.1~25.3°C. The environment in the classroom could not meet students' demands for thermal comfort and air quality. Students' adaptive behaviors included adding or reducing clothes, light activities, opening doors and windows for ventilation, etc.

Keywords Cold region; Transitional season; Thermal comfort; Air quality

蒋婧，西安工程大学，副教授，硕士生导师。
刘菲，西安工程大学，硕士研究生。
云旭鑫，西安工程大学，硕士研究生。

居民活动与住区微气候的时空耦合对长期热舒适的影响

Influence of Spatial and Temporal Coupling of Occupants' Activity and Residential Microclimate on Residents' Thermal Comfort

刘益，宋德萱，余翔宇，王正阳，于卓玉
Liu Yi, Song Dexuan, Yu Xiangyu, Wang Zhengyang, Yu Zhuoyu

摘要 住区是居民日常活动的主要场所，合理的住区热环境不仅是居民健康的基本保障。目前，有许多学者基于热舒适理论对住区微气候及其物质环境进行了研究，但大部分研究忽略了不同时间、不同空间的居民活动对热舒适的影响。本文对上海市 2 个住宅小区进行了现场调研与环境模拟，发现了住区之间居民活动与微气候在时空耦合上的差异性。同时，基于对 122 位居民的问卷调研，分析了居民活动行为与微气候的时空耦合对居民长期热舒适的影响，为更精准有效地评估住区室外热舒适提供了新的方向。

关键词 居民热舒适；住区微气候；居民活动；时空耦合

Abstract The neighborhood is the primary place for the daily activities of the residents. A reasonable thermal environment in a neighborhood is a basic guarantee for the health of the residents. At present, many scholars have studied the microclimate of the neighborhood and its physical environment for residents' thermal comfort, but most of them have ignored that how residents' activities at different times and in different spaces affect their thermal comfort. In this paper, we conducted field studies and environmental simulations in two residential areas in Shanghai, and found that the spatio-temporal coupling of residents' activities and microclimate differed between residential areas. The impact of this spatio-temporal coupling on residents' thermal comfort was then analysed based on a questionnaire survey of 122 residents, which providing a new direction for a more accurate and effective assessment of outdoor thermal comfort in residential areas.

Keywords Residents' thermal comfort; Microclimate of residential area; Residents' activities; Spatio-temporal coupling

刘益，余翔宇，王正阳，于卓玉，同济大学建筑与城市规划学院，博士研究生。
宋德萱，同济大学建筑与城市规划学院，教授，博士生导师。

基于风热耦合模拟方法的湿热地区半室外空间吊扇布局研究

Research on Layout of Ceiling Fans in a Semi-outdoor Space in Hot and Humid Area Based on the Coupled Simulation Method of Wind and Thermal Environment

肖毅强，胡婉婷，吕瑶，林瀚坤 *
Xiao Yiqiang, Hu Wanting, Lü Yao, Lin Hankun

肖毅强，华南理工大学建筑学院，教授，设计学院院长。
胡婉婷，华南理工大学建筑学院。
吕瑶，华南理工大学建筑学院，副教授。
林瀚坤，广东工业大学建筑与城市规划学院，助理研究员。

摘要 在湿热地区，吊扇是一种使用广泛、能耗较低的主动式气流调节措施，能有效地改善热环境。本研究以广州某半室外空间吊扇（直径 1.4 m）为研究对象，在过渡季典型日实测了温湿度、黑球温度和风速等热环境指标。进而，通过 Fluent 软件对吊扇不同布局进行模拟，并结合 Ladybug Tools 软件耦合计算不同工况下的热舒适指标 PET。研究结果表明，吊扇中档位运行时平均风速为 0.6 m/s，PET 值比自然通风（无风扇）工况低 1.5℃，比静风（无风扇）工况低 4℃。吊扇中心距离 3.5 m 布局比距离 5.6 m 的 PET 值降低 0.3℃，舒适区占比增加了 25.5%。

关键词 湿热地区；半室外空间；吊扇；热舒适；CFD；耦合模拟方法

Abstract As an active flow regulator with low energy consumption, ceiling fans can effectively improve the thermal environment. The semi-outdoor space ceiling fans (1.4 m in diameter) in Guangzhou was taken as an example. The thermal environment indicators of a typical day during the transitional season were measured, including temperature, humidity, globe temperature, and wind speed. Fluent simulations were performed on different fan layouts, and the Ladybug Tools calculated the Physiological Equivalent Temperature (PET) comfort evaluation index under different conditions. The results shows with an average wind speed is 0.6m/s, with PET values 1.5°C lower than the natural ventilation (no fan) scenarios and 4°C lower than still air (no fan) scenarios when the ceiling fan is running in the mid range. A fan layout with a 3.5 m center distance reduces the PET value by 0.3°C compared to a 5.6 m distance, increasing the comfort zone percentage by 25.5%.

Keywords Hot and humid area; Semi-outdoor space; Ceiling fans; Thermal comfort; CFD; Couple simulation method

---

汉江中游传统民居典型空间的夏季热环境分析

Analysis of the Summer Thermal Environment of a Typical Space in Traditional Houses in the Middle Reaches of the Han River

刘越，杨柳，袁玮晴，陈飞宇，刘衍
Liu Yue,Yang Liu,Yuan WeiQing,Chen FeiYu,Liu Yan

刘越，西安建筑科技大学建筑学院，博士研究生；湖北文理学院土木工程与建筑学院，讲师。
杨柳，西安建筑科技大学建筑学院，教授，博士生导师。
袁玮晴，西安建筑科技大学建筑学院，博士研究生。
陈飞宇，西安建筑科技大学建筑学院，硕士研究生。
刘衍，西安建筑科技大学建筑学院，教授，博士生导师。

摘要 选取汉江中游十堰地区传统民居饶氏庄园为研究对象，针对其典型空间进行夏季热环境现场测试分析，进一步计算平均辐射温度、体感温度与露点温度等参数作对比研究，结果表明不同空间类型呈现出显著的热环境差异性；合理的空间组织与布局、适应气候的遮阳与通风策略、围护结构材料性能均能起到关键的调适作用。本文以挖掘传统民居生态智慧经验为出发点，为该地区新建居住建筑的节能设计及改善室内热环境提供参考依据作出贡献。

关键词 传统民居；典型空间；夏季热环境；现场测试

Abstract Raao Manor, a traditional residence in the Shiyan area in the middle reaches of the Han River, was selected as the research object, and its typical space was tested and analysed in the summer thermal environment. The average radiation temperature, body temperature, dew point temperature and other parameters were calculated for comparative analysis, and it was found that different space types showed significant differences in thermal environment. Appropriate spatial organization and layout, climate-adapted shading and ventilation strategies, and material properties of the enclosure structure can all play a key role in adaptation. This paper takes the experience of traditional residential ecological wisdom as the starting point to provide reference for the energy-saving design and improvement of indoor thermal environment of new residential buildings in this area in the later period.

Keywords Traditional dwellings; Typical space; Thermal environment in summer; Field testing

过渡季整体与身体局部热感觉相关性研究

A Study of the Correlation Between Whole Body and Local Body Thermal Sensation in Transitional Season

赵奂琛，夏博，张新雨，严舸飞
Zhao Huanchen, Xia Bo, Zhang Xinyu, Yan Gefei

赵奂琛，长安大学建筑学院，硕士研究生。
夏博，长安大学建筑学院，副教授。
张新雨，长安大学建筑学院，硕士研究生。
严舸飞，长安大学建筑学院，硕士研究生。

摘要 为研究人的整体热感觉与身体局部热感觉之间的相关性，以西北高校学生为研究对象。本文选取寒冷地区，北方西安的一所大学建筑进行了实地研究与问卷调查。通过物理环境测试和受试者的主观问卷调查，对比人整体与身体各部位之间的热感觉差异；结果表明，在多数情况下，热感觉在人体各个部位均有差异，人体的胸部的热感觉与整体热感觉最为接近。人体各部位对于整体热感觉的影响权重从大到小分别为：头颈部，脚部，腿部，胸部，背部，肩部。

关键词 西北高校；整体热感觉；局部热感觉；相关性分析

Abstract To explore the correlation between overall and localized thermal sensation in humans, a study was conducted using students from a Northwestern university as research subjects. The research took place in a university building in the cold region of Xi'an in northern China. It included field research and questionnaire surveys. By combining physical environmental tests and subjective questionnaires with the participants, the study compared the differences in thermal sensation between the whole body and different parts of the body. The results showed that, in most cases, thermal sensation differs between different parts of the body, with the chest area being closest to the overall thermal sensation. In addition, the weighted influence of different parts the body on overall thermal sensation was ranked from highest to lowest as follows: head and neck, feet, legs, chest, back, and shoulders.

Keywords Northwestern universities; Overall thermal sensation; Local thermal sensation; Correlation analysis

---

风速及风向对服装整体和局部热阻影响实验研究

Experimental Investigation into the Effects of Wind Direction and Speed on Total and Local Clothing Insulation

高姗，Ooka Ryozo，Oh Wonseok，冯驰
Gao Shan, Ooka Ryozo, Oh Wonseok, Feng Chi

高姗，重庆大学建筑城规学院，助理研究员。
Ooka Ryozo，东京大学生产技术研究所，教授。
Oh Wonseok，东京大学生产技术研究所，助教。
冯驰，重庆大学建筑城规学院，教授，建筑技术科学系副系主任。

摘要 作为多节点体温调节模型的必要输入参数，目前局部服装热阻的数据较为缺乏，多数研究采用整体热阻代替人体各部位的局部服装热阻，制约了体温调节模型的预测精度。本研究采用 16 区段暖体假人，在边界层风洞中，测试 7 套日常服装的整体和局部热阻。研究发现站姿服装整体热阻高于坐姿服装整体热阻，且侧向气流工况下服装整体热阻最高；胸部、背部以及腰腹部的服装热阻高于其他部位。基于测试结果，本文提出了不同风速条件下整体和局部服装热阻预测公式。本研究结论有助于完善服装热阻数据库及提高体温调节模型的预测精度。

关键词 服装热阻；暖体假人；风速；风向；热舒适

Abstract Current research on local clothing insulation lags compared with multi-node thermoregulation models, which affectsmodel prediction accuracy. Therefore, in this study, the total and local clothing insulation values for 7 sets of typical clothing ensembles were determined based on thermal manikin experiments. The results showed that the total clothing insulation for the standing posture was larger than that for the sitting posture, and the total clothing insulation caused by the crosswind was the highest. And the local clothing insulations for the chest, back, and pelvis were larger higher than those of other body segments. The prediction equations for total and local clothing insulation under different airspeeds were proposed based on the results of the experiment. The results of this study help to improve the clothing insulation database and the prediction accuracy of thermoregulation models.

Keywords Clothing insulation; Thermal manikin; Airspeed; Wind direction; Thermal comfort

不同短期热经历对人体热适应的影响研究

Influences of Different Short-term Thermal Histories on Human Adaptive Thermal Comfort

潘羽，石芳宁，孙震，闫海燕 *
Pan Yu, Shi Fangning, Sun Zhen, Yan Haiyan

摘要 短期热经历对人体热舒适的研究对于优化公共建筑夏季空调节能具有重要意义。本文采取实验研究的方法，选取焦作市某高校一装有分体式空调的房间，探讨了由26℃和20℃设定温度引起的不同短期热经历对人体热适应的影响。结果表明，在相同操作温度下，相比于20℃，26℃工况下受试者的皮肤温度更低，皮肤温度对环境温度的敏感性更高。26℃工况下受试者的热感觉在空调开启后40 min进入稳定状态，20℃工况下受试者的热感觉一直在降低，但随着时间的推移，显著降低受试者热感觉的时间越来越长。

关键词 短期热经历；人体适应性；热舒适；热反应

Abstract The study of short-term thermal experience on human thermal comfort is of great significance for optimizing summer air conditioning energy efficiency in public buildings. This article adopts an experimental research method and selects a room equipped with split air conditioning in a university in Jiaozuo to explore the impact of different short-term thermal experiences caused by temperature settings of 26°C and 20°C on human thermal adaptation. The results showed that under the same operating temperature, the skin temperature of the subjects was lower compared to 20°C and 26°C, and the sensitivity of skin temperature to environmental temperature was higher. The thermal sensation of the subjects under 26°C operating conditions remained stable for 40 minutes after the air conditioning was turned on. At 20°C, the thermal sensation of the subjects continued to decrease, but over time, the time that significantly reduced the thermal sensation of the subjects became longer and longer.

Keywords Short-term thermal experience; Human adaptability; Thermal comfort; Thermal reaction

潘羽，河南理工大学，硕士研究生。
石芳宁，河南理工大学，硕士研究生。
孙震，河南理工大学，硕士研究生。
闫海燕，河南理工大学，教授。

季节变化和天气状况对建筑垂直绿化热效应的影响

Influence of Seasonal Variation and Weather Condition on Thermal Effect of Vertical Greening

姜之点，杨峰 *
Jiang Zhidian, Yang Feng

摘要 垂直绿化热效应的季节性差异以及阴、雨等天气状况可能引发的负向效应较少被讨论。在夏热冬冷地区的上海，对覆有常绿垂直绿化的办公建筑室外热环境开展长时间序列观测实验。年度观测结果显示，垂直绿化整体表现为降温效应，但降温天数不足全年一半。南向垂直绿化较北向具有更高热效应强度，两朝向平均降温强度相差较小。在理想气象条件下，南向垂直绿化四季均以降温为主，而北向降温仅集中在夏、秋两季。垂直绿化降温强度在晴朗天气下最大，随降雨量增加而减小，随风力等级增加而递增。研究结果将有助于加深对垂直绿化长效机制的全面理解。

关键词 垂直绿化；季节变化；天气状况；长时间序列；热效应

Abstract The possible negative effects of vertical greening system (VGS) have received less attention under different seasonal changes and weather conditions (e.g. cloudy and rainy). In hot summer and cold winter zones, long-term series observation experiments were conducted on public buildings covered with evergreen VGS. The annual observation results showed that VGS generally exhibited a cooling effect, but the cooling days were less than half a year. South-facing VGS had higher intensity of thermal effects than north-facing, but the average cooling intensity of the two differed little. Under ideal weather conditions, south-facing VGS was dominated by cooling in four seasons, while north-facing cooling only concentrated in summer and autumn. The cooling intensity of VGS was the greatest on sunny days, decreased with increasing rainfall, and increased with increasing wind speed. The findings will contribute to a dialectical understanding of the long-term mechanism of VGS.

Keywords Vertical greening system; Seasonal variation; Weather condition; Long-term series; Thermal effect

姜之点，同济大学建筑与城市规划学院，博士研究生。
杨峰，同济大学建筑与城市规划学院，博士，博士生导师，副教授。

以性能为导向的建筑自然通风生成设计方法

Performance-oriented Building Natural Ventilation Generative Design

肖钰澄，庄智*，张菁燕，赵田，王沁芳

Xiao Yucheng, Zhuang Zhi, Zhang Jingyan, Zhao Tian, Wang Qinfang

肖钰澄，同济大学建筑与城市规划学院，硕士研究生。

庄智，同济大学建筑与城市规划学院/同济大学中德工程学院，副教授。

张菁燕，常州市建筑科学研究院，总工程师。

赵田，同济大学建筑与城市规划学院，博士研究生。

王沁芳，常州市建筑科学研究院，主任工程师。

摘要 在“双碳”目标下建筑节能设计要求越来越高，在适宜气候条件下充分利用自然通风可有效地提高室内舒适度，并减少建筑空调系统使用时间及运行能耗。以自然通风性能为导向的生成设计有助于指导建筑师科学地开展方案设计，进而提高建筑自然通风的利用率。本文提出了以窗墙比、窗户宽高比等参数为通风设计变量，自然室温舒适小时数与建筑冷负荷为优化目标函数，利用遗传算法得到设计变量的帕累托前沿解，最后基于熵权法得到最优方案的自然通风生成设计方法，为建筑自然通风性能生成设计提供思路。

关键词 性能导向；生成设计；自然通风；公共建筑

Abstract Under the carbon peaking and carbon neutrality goals, the energy-saving design requirements of buildings are getting higher and higher. Making full use of natural ventilation under suitable climatic conditions can effectively improve indoor comfort and reduce the use time and energy consumption of air conditioning systems. Conducting a performance-oriented generative design can help guide architects to carry out scheme design scientifically, which can improve the natural ventilation potential in buildings. This paper the building natural ventilation generative design is proposed, including the ventilation design variables such as window wall ratio and window width-height ratio, the optimization target function of the natural room temperature comfortable hours and the building cold load, the Pareto frontier solutions of the design variable obtained by the genetic algorithm, and the optimal solution based on the entropy weight method. This study provides ideas for the natural ventilation performance generative design of the building.

Keywords Performance-oriented; Generative design; Natural ventilation; Public buildings

---

应用预计冷热感指标探寻室内制冷环境的至适温度

Research on Comfort Threshold Temperature of the Indoor Cooling Environment Based on the Predicted Mean Vote

郑华美

Zheng Huamei

郑华美，福州大学建筑与城乡规划学院，讲师，博士。

摘要 本研究基于PMV的热舒适方程式计算方法，在热舒适指标0≤PMV≤1，以及夏季热舒适设计参数范围内，算得空气温度等于平均辐射温度的温度阈值。在居室者夏季常规着装量和代谢量状态下的温度阈值范围内，得出室内制冷环境的至适温度。在相对湿度≤70%、气流速度≤0.25 m/s的条件下，PMV≤1、PMV≤0.5的至适温度分别为27.3~28.2 ℃、25.8~26.8℃。满足PMV=0的相对湿度40%、气流速度≤0.25 m/s的至适温度为24.8~26.0℃。

关键词 PMV；制冷环境；温度阈值；至适温度

Abstract This paper of object is to research on comfort threshold temperature for the indoor cooling environment in summer. According to the thermal comfort index of 0≤PMV≤1 and the parameters of thermal comfort design, the temperature threshold has been calculated based on the thermal comfort equations of PMV while the mean radiant temperature is equal to the air temperature. Under the range of temperature threshold of the normal metabolic rate and clothing insulation in summer, the comfort threshold temperatures have been obtained in the indoor cooling environment. The range of comfort threshold temperature have been determined 27.3~28.2°C for PMV ≤ 1, 25.8~26.8°C for PMV ≤ 0.5 when the relative humidity ≤ 70% and relative velocity ≤ 0.25 m/s. The range of comfort threshold temperature is 24.8~26.0°C for PMV=0 when the relative humidity 40% and relative velocity ≤ 0.25 m/s.

Keywords Predicted Mean Vote; Cooling environment; Temperature threshold; Comfort threshold temperature

晋北农村分散采暖建筑室内热环境实测研究

Experimental Study on Indoor Thermal Environment of Distributed Heating Buildings in Rural Areas of Northern Shanxi Province

冯晶，董旭娟，张孟莹
Feng Jing, Dong Xujuan, Zhang Mengying

冯晶，太原理工大学，硕士研究生。
董旭娟，太原理工大学，副教授，硕士生导师。
张孟莹，太原理工大学，硕士研究生。

摘要 以晋北地区农村单体建筑为研究对象，采用实地测试结合问卷调查的方法，对山西省忻州市河曲县岱岳殿村的 4 处使用不同分散采暖方式的建筑室内热环境进行了测试研究。结果表明：空气能热泵结合暖气片采暖方式下的建筑室内热环境能够满足居民的热舒适需求，且已在全村普遍应用，此外，采暖末端为地暖对室内空气温度提升效果明显，为农村既有建筑更换采暖方式及新建建筑选取采暖方式提供选择。

关键词 晋北地区；农村单体建筑；分散采暖；室内热环境

Abstract Taking rural single buildings in the northern region of Shanxi Province as the research object, the indoor thermal environment of four buildings using different decentralized heating methods in Daiyuedian Village, Hequ County, Xinzhou, Shanxi Province was tested and studied using field testing and questionnaire survey methods. The results show that the indoor thermal environment of buildings under the air energy heat pump combined with radiator heating mode can meet the thermal comfort needs of residents, and has been widely used in the whole village. In addition, the floor heating at the heating end has an obvious effect on the indoor air temperature increase, providing options for the replacement of heating methods for existing buildings in rural areas and the selection of heating methods for new buildings.

Keywords North Shanxi region; Rural single building; Distributed heating; Indoor thermal environment

---

基于遗传算法的建筑 RC 模型辨识及室内温度预测研究

Research on Building RC Model Identification and Indoor Temperature Prediction Based on Genetic Algorithm

张欣林，陈源昕，李怀，郝玉珍，曲明璐
Zhang Xinlin, Chen Yuanxin, Li Huai, Hao Yuzhen, Qu Minglu

张欣林，上海理工大学环境与建筑学院，硕士研究生。
陈源昕，南京工业大学城市建设学院，硕士研究生。
李怀，中国建筑科学研究院，正高工程师。
郝玉珍，中国建筑科学研究院，初级工程师。
曲明璐，上海理工大学环境与建筑学院，副教授。

摘要 为了准确预测建筑物室内温度并获得最佳的控制策略，本文在采用集总参数法对问题进行简化的基础上，设计建筑热阻容（RC）模型，并推导出相应的微分方程组，使用遗传算法对模型进行辨识。RC 模型测试集温度平均绝对误差为 0.14℃，R^2 为 0.99。本文还讨论了黑箱模型和灰箱模型的适用条件，根据实际问题的需求和数据情况，综合考虑模型的复杂度、预测精度、解释性等因素，选择合适的建模方法。

关键词 遗传算法；温度预测；热阻容模型；黑箱模型；建筑节能

Abstract In order to accurately predict the indoor temperature of buildings and obtain the optimal control strategy, this paper simplifies the problem using the lumped parameter method, designs a building thermal resistance and capacitance (RC) model, and derives the corresponding differential equation system. The model is identified using a genetic algorithm. The average absolute error of the RC model test set temperature is 0.14°C, and the R^2 is 0.99. This paper also discusses the applicable conditions of black box and grey box models, and selects the appropriate modeling method based on the complexity, prediction accuracy, interpretability, and other factors of the model, considering the actual requirements and data situation of the problem.

Keywords Genetic algorithm; Temperature prediction; RC model; Black box model; Building energy conservation

我国农村居住建筑热环境研究趋势可视化分析

Visualization Analysis of Thermal Environment Research Trend of Rural Residential Buildings

刘欣蕊，陈俊，王佳婧，杨文健，王岩 *

Liu Xinrui, Chen Jun, Wang Jiajing, Yang Wenjian, Wang Yan

刘欣蕊，陈俊，王佳婧，天津城建大学建筑学院，硕士研究生。
杨文健，天津城建大学国际工程学院，硕士研究生。
王岩，天津城建大学建筑学院，工学博士，教授。

摘要 通过整理归纳我国农村居住建筑热环境的研究趋势，为今后相关研究的深入开展提供科学参考依据。本文采用文献计量法，对“中国知网”数据库中 2003 至 2023 年的 849 篇相关文献进行统计分析。结果发现近 20 年我国农村居住热环境研究经历了起始阶段—发展阶段—高潮阶段；研究热点为“农村住宅”“热舒适”“热环境”“围护结构”等高频关键词。“节能改造”“农宅”“优化策略”“人居环境”正处于突现期。在新的政策引导下，运用现代技术对农村居住建筑的热环境营造进行优化提升和节能改造是未来的发展趋势。

关键词 农村住宅；热环境；文献计量软件；图谱分析

Abstract Research trends in the thermal environment of residential buildings in rural China are summarized to provide a scientific reference for future research. In this paper, we used biometric method to analyze 849 relevant literature from 2003 to 2023 in the database of “China Knowledge Network”. It was found that over the past 20 years, research on the thermal environment of rural housing in China has gone through an initial phase, a development phase and a climax phase; the research hotspots are “rural housing”, “thermal comfort” and “thermal environment”. “envelope” are high-frequency keywords. “Energy-saving renovation”, “rural housing”, “optimization strategy” and “habitat environment” are in the emergence period. Under the guidance of the new policy, the future development trend is to optimize and improve the thermal environment of rural residential buildings with modern technology and energy-saving renovation.

Keywords Rural residence; Thermal environment; CiteSpace; Knowledge map

---

抽油烟机排风方式与位置对厨房气流组织的影响研究

Research on the Influence of Exhaust Mode and Position of Range Hood on the Indoor Air Flow Organization of Kitchen

廖淑琦，闫增峰，倪平安，雷馥铭

Liao Shuqi, Yan Zengfeng, Ni Pingan, Lei Fuming

廖淑琦，西安建筑科技大学，硕士研究生。
闫增峰，西安建筑科技大学，教授。
倪平安，西安建筑科技大学，博士研究生。
雷馥铭，西安建筑科技大学，硕士研究生。

摘要 厨房作为住宅中的重要功能空间，其室内热环境及空气质量与人体健康息息相关。本研究以西安市某住宅厨房为对象，运用数值对夏季的厨房气流组织进行模拟研究，分析讨论厨房设备及布局对室内热环境和空气质量的影响，结果表明：在窗户邻侧时，顶吸式抽油烟机的人体面部温度更低，侧吸式抽油烟机的室内二氧化碳浓度最低。

关键词 住宅厨房；室内热环境；气流组织；数值模拟

Abstract Kitchen is an important functional space in the house, and its indoor thermal environment and air quality are closely related to human health. In this study, a residential kitchen in Xi 'an was used as an object to simulate the kitchen air flow organization in summer by numerical simulation, and the influence of kitchen equipment and layout on indoor thermal environment and air quality was analyzed and discussed. The results showed that when the window was adjacent, the human face temperature of the top suction range hood was lower, and the indoor carbon dioxide concentration of the side suction range hood was the lowest.

Keywords Residential kitchen; Indoor thermal environment; Air flow structure; Numerical simulation

自然通风模式对皖南某居住建筑热环境的影响探究

Study of Natural Ventilation Mode Effect on a Residential Indoor Thermal Comfort in South Area of Anhui Province

陈万鹏，郭方棚 *
Wanpeng Chen,
Fangpeng Guo

摘要 随着“双碳”目标的提出，节能形势愈发严峻。为探究自然通风条件下住宅建筑的节能潜力，本文以芜湖市某居住建筑作为研究对象，利用建筑性能分析软件 Ladybug Tools 探究不同自然通风模式（开窗面积、开窗时间等）对室内热环境的影响。研究表明：开窗面积比小于 0.5 时，室内的舒适小时数随着开窗面积的增大而显著增加；面积比达到 0.5 后，继续增加，对舒适小时数的提升不明显。就基于温度控制设置的通风模式而言，温度区间 15~30℃比温度区间 18~26℃更有利于提高室内的舒适小时数。

关键词 居住建筑；开窗面积；室内热环境；建筑模拟

Abstract With the proposal of carbon peaking and carbon neutrality goals, the form of energy saving is becoming more and more serious. In order to explore the energy saving potential of residential buildings under natural ventilation conditions, this paper takes a residential building in Wuhu as the research object and uses Ladybug Tools, a building performance analysis software, to explore the impact of natural ventilation modes (window opening area, window opening time, etc.) on indoor thermal environment. The results show that when the window opening area ratio is less than 0.5, the indoor comfort hours increase significantly with the increase of window opening area. After the area ratio reaches 0.5, it continues to increase, and the improvement of comfort hours is not obvious. For the ventilation mode based on the temperature control setting, the temperature interval of 15~30°C is more conducive to improving the indoor comfort hours than the temperature interval of 18~26°C.

Keywords Residential buildings; Window area; Indoor thermal environment; Building simulation

陈万鹏，安徽工程大学建筑工程学院，本科生。
郭方棚，安徽工程大学建筑工程学院，博士，讲师。

基于 GB sware 模拟的既有建筑风热环境优化设计

Wind and Thermal Environment Optimization Design of Existing Buildings Based on GB Sware Simulation

姜玲，黄海静，张福超
Jiang Ling, Huang Haijing, Zhang Fuchao

摘要 建筑模拟软件由于其环境因素可视化呈现的特点，对指导建筑设计方案起着十分重要的作用。本文结合 2021 年绿建斯维尔竞赛特等奖参赛作品，采用绿建斯维尔 GB sware 软件，通过对建筑室内外风热环境的模拟分析，提出既有建筑的绿色改造策略及建筑风热环境优化设计。该研究显示，改造后的建筑能耗大幅下降，基础节能率达到 86.56%，对比节能率达到 61.61%，建筑全生命周期碳排放实现近零能耗建筑标准，并且负碳。

关键词 GB sware；环境模拟；绿色建筑；节能改造策略

Abstract Building simulation software plays a very important role in guiding architectural design due to its visual presentation of environmental factors. In this paper, combined with the 2021 first prize entries of Green Building Competition, using GB sware software, through the simulation analysis of the indoor and outdoor wind and thermal environment of the building, the green transformation strategies of the existing building and the optimization design of the wind and thermal environment of the building are proposed. The research shows that the energy consumption of the building after the transformation is greatly reduced, the basic energy saving rate reaches 86.56 %, and the comparative energy saving rate reaches 61.61 %. The carbon emission of the whole life cycle of the building achieves the near-zero energy consumption building standard, even negative carbon.

Keywords GB sware; Environmental simulation; Green building; Energy saving strategy

姜玲，重庆大学建筑城规学院，硕士研究生。
黄海静，重庆大学建筑城规学院，山地城镇建设与新技术教育部重点实验室，教授。
张福超，重庆大学建筑城规学院，硕士研究生。

第 5 篇

建筑门窗与幕墙

基于 THERM 模拟研究穿透隔热层的螺栓对玻璃幕墙热工性能的影响

Impact of Bolt Penetrating the Insulation Layer on the Curtain-Wall Thermal Performance Based on THERM Simulation

周雨祯 *
Zhou Yuzhen

摘要 隔热断桥铝框成为现代玻璃幕墙的“标配”，然而断桥隔热条的材质大多为有机聚合物，为解决其作为结构传力构件耐久性差的隐忧，工程设计使用螺栓直接连接室内和室外的金属框架。为了研究这些穿透隔热层螺栓对玻璃幕墙热工性能的影响，本文将多种直径、间距的碳钢和不锈钢螺栓与多种隔热断桥铝框进行组合，使用 THERM 热工仿真软件模拟框架截面，再根据“平行截面”和“等效导热”两种理论算法进行框和幕墙整体传热系数的综合计算。研究结果显示：碳钢和不锈钢螺栓分别使框传热系数增加 84% 和 42%、使整体幕墙的传热系数增加 23% 和 12%，其影响不可忽略。

关键词 玻璃幕墙；幕墙热工设计；非通长热桥构件；U 值

Abstract Thermally-broken aluminum frames become the “mandatory componen” for modern curtain walls. However, most of the thermal break materials are organic polymers, so as to solve the concern of poor durability as a structural force transmission component; bolts are designed to directly fix indoor and outdoor metal frames. In order to study the actual impact of bolts penetrating the thermal insulation layer, this paper combines carbon steel and stainless-steel bolts with various diameters and spacing with a variety of thermally-broken extrusions, uses THERM software to simulate the frame cross-section, and then comprehensively calculates the overall heat transfer coefficient of the frame and curtain wall according to the two theories of “parallel cross-section” and “effective thermal conductivity”. The results show that carbon steel and stainless-steel bolts increase the heat transfer coefficient of the frame by 84% and 42%, and the overall heat transfer coefficient of the curtain wall by 23% and 12%. The thermal bridging effect of the bolts should not be negligible.

Keywords Curtain wall; Facade thermal design; Non-continuous thermal bridging element; U-value

周雨祯，暖框科技公司技术总监，同济大学土木工程学士，美国 NFRC 认证热工建模师，美国 LEED 绿建专才。

彩色半透明光伏组件的光性能参数设计及实测评价

Parameter Design and Experimental Evaluation of the Optical Performance of Colored Semi-transparent PV Modules

师劭航，朱宁，宋晔皓
Shi Shaohang, Zhu Ning, Song Yehao

摘要 建筑光伏一体化组件的光性能设计影响了建筑能效和视觉效果。本研究以无色相和彩色透光碲化镉光伏窗为例，测试了晴天室内外光环境。通过照度、显色性等分析，评价了 15 个光伏窗的透射光，并分析了参数设计对其光性能的影响。结果表明，对于同色相的组件，随着光伏覆盖率增加，室内光照度呈现非线性增加，但光色无明显差异。不同色相组件的色彩逼真度指数（Rf）和色彩饱和度指数（Rg）的最大差值分别为 88.74 和 136.3，显色指数（Ra）排序为橙 > 黄 > 蓝 > 红。结合围护结构光性能的设计需求，对彩色半透明光伏组件的参数设计提出了优化建议。

关键词 建筑光伏一体化，彩色光伏，透光围护结构，日光，显色性

Abstract The optical performance design of building-integrated photovoltaic modules determines the energy efficiency and visual effects of buildings. In this study, indoor and outdoor lighting environments on sunny days were tested with colorless and colored semi-transparent CdTe PV windows as an example. The transmitted light of 15 PV windows was evaluated by the analysis of illuminance and color rendering, and the parameters designed on their optical performance were analyzed. The results show that for the modules with the same color, the indoor illumination shows a non-linear increase with the increase of PV coverage ratios, but there is no significant difference in light color. The maximum differences of color fidelity index (Rf) and gamut index (Rg) for modules of different hues are respectively 88.74 and 136.3, and the color rendering index (Ra) is ranked as orange > yellow > blue > red. Optimization suggestions for the parameter design of colored semi-transparent PV modules are provided for building envelope.

Keywords Building-integrated photovoltaic; Colored PV module; Transparent envelope; Daylighting; Color rendering

师劭航，清华大学建筑学院，博士研究生。
朱宁，清华大学建筑学院 / 清华大学生态规划与绿色建筑教育部重点实验室，副教授。
宋晔皓，清华大学建筑学院 / 清华大学生态规划与绿色建筑教育部重点实验室，教授，建筑与技术研究所所长。

盐雾气候下建筑玻璃表面盐分沉积特性的实验研究

Experimental Study on Salt Deposition Characteristics of Building Glass Surface in Salt Fog Climate

毛会军，孟庆林
Mao Huijun, Meng Qinglin

摘要 以市场上常见的中空 Low-E 玻璃为研究对象，通过加速盐雾试验方法使玻璃表面产生盐分沉积，探究了不同喷雾时间、盐水浓度与沉积特性之间的关系。结果表明，随着喷雾时间变长，玻璃表面的盐斑分布均匀性与密集程度均有所下降，但盐斑尺寸与喷雾时间呈现指数函数增长趋势，而单位面积沉积量与喷雾时间呈现对数函数关系；高盐水浓度下玻璃试件表面单位面积沉积量与喷雾时间仍呈现对数增长趋势，但高浓度对应的单位面积沉积量始终大于低浓度工况，其平均增长幅度约为 11.6%。该发现可为进一步研究盐雾气候下建筑玻璃热工性能变化规律奠定基础。

关键词 盐雾气候；建筑玻璃；加速盐雾实验；盐分沉积；喷雾时间；盐水浓度

Abstract In this paper, the salt deposition on the hollow Low-E glass surface was first yielded by the method of accelerated salt spray test, and the relationship between spray time, salt concentration and deposition characteristics was explored. The results show that the distribution uniformity and density of salt spots on the glass surface decrease as spray time increases, but the size of salt spots grows exponentially with increasing spray time. However, the deposition amount per unit area of the glass surface increases logarithmically with a higher spray time. The deposition amount and spray time also show logarithmic growth even at high saline concentration, but the deposition amount at high saline concentration is always higher than that at low saline concentration, with an average increase of 11.6%. This finding can lay a foundation for further research on how the thermal performance of building glass varies in salt fog climate.

Keywords Salt fog climate; Building glass; Accelerated salt spray test; Salt deposition; Spray time; Saline concentration

毛会军，华南理工大学建筑学院，博士研究生。
孟庆林，华南理工大学建筑学院，教授，亚热带建筑科学国家重点实验室副主任，中国建筑学会建筑物理分会理事长。

一种建筑门窗幕墙用保温与遮阳性能可调卷膜系统的节能效果模拟研究

Simulation Study on Energy Saving Effect of a Performance Adjustable Film System for Building Windows, Doors, and Curtain Walls

刘志超，孟庆林，毛会军
Liu Zhichao, Meng Qinglin, Mao Huijun

摘要 门窗幕墙对建筑能耗的影响极大，性能可调围护结构可进一步提高建筑的气候适应性与节能效果。本文提出了一种建筑门窗幕墙用保温与遮阳性能可调卷膜系统，模拟研究了其搭配不同玻璃配置时对建筑能耗的影响，并与传统内遮阳卷帘的节能效果进行了对比。结果表明，当以立面接收到的太阳直射辐射强度作为控制指标进行动态调节，玻璃热工性能较高与适中时，该新型遮阳系统在各热工分区产生的最大节能率分别为 16.4% 与 8.5%，显著高于搭配传统卷帘的遮阳系统。该新型遮阳系统可以低成本的方式进一步降低建筑能耗。

关键词 门窗幕墙；保温与遮阳；性能可调；卷膜系统；建筑能耗模拟

Abstract The building's energy consumption is greatly influenced by the windows, doors, and curtain walls, and performance adjustable building envelope can improve the building's climate adaptability and energy-saving capabilities. In this paper, a performance adjustable film system for building windows, doors and curtain walls was proposed, the influence of different glass configurations on building energy consumption is simulated, and the energy-saving effect is compared with that of a traditional internal shading screen. When using the direct solar radiation intensity received by the facade as the control index for dynamic adjustment, the results show that the novel shading system can achieve the maximum energy-saving rates of 16.4% and 8.5%, under high and moderate glass thermal performance respectively, significantly higher than the traditional internal shading system. The novel shading system provides a low-cost way to further reduce building energy consumption.

Keywords Building windows, doors and curtain walls; Insulation and shading; Performance adjustable; Film system; Building energy simulation

刘志超，华南理工大学建筑学院，硕士研究生。
孟庆林，华南理工大学建筑学院教授，亚热带建筑科学国家重点实验室副主任，中国建筑学会建筑物理分会理事长。
毛会军，华南理工大学建筑学院，博士研究生。

建筑集成聚光百叶对室内照度分布的影响研究

Study on Influence in Indoor Illumination Distribution for Building Integrated Solar-Concentrated Shutters

邵泽彪，蔡吕培 *
Shao Zebiao, Cai Lüpei

摘要 建筑集成透射式聚光百叶，可将直射光用于产能、散射光用于采光，实现对于天然光线高效的层级利用。本文模拟研究了集成太阳能聚光百叶在不同工况下的室内照度分布，并与普通玻璃窗及集成普通百叶窗口进行了对比。研究结果表明：相比普通玻璃窗，当集成聚光百叶处于工作状态时，室内进光量减少，在聚光百叶兼具遮阳作用的同时，房间内的光线分布更加均匀；相比集成普通百叶窗口，集成聚光百叶窗口进入室内的天然光线更多，且在提高房间深处照度方面也更具优势。

关键词 聚光百叶；天然光线；照度分布；室内光环境

Abstract The building integrated transmissive solar-concentrated shutters can use direct light for energy production and scattered light for daylighting, realizing efficient hierarchical utilization of sunlight. The indoor illumination distribution of integrated solar-concentrated shutters under different working conditions was simulated and compared with conventional glass windows and integrated conventional shutters windows. The results of the study show that, compared to regular glass windows, the amount of light entering the room was reduced when the integrated concentrating shutters were in working condition. This indicates that the concentrated shutters have a shading effect and the light distribution in the room with integrated concentrated louvers is more uniform. Compared with integrated regular shutters, windows integrated concentrated shutters have more natural light entering the room and are also more advantageous in increasing the illumination in the depth of the room.

Keywords Solar-concentrated shutters; Daylighting; Illumination distribution; Indoor light environment

邵泽彪，中国矿业大学，博士，讲师，硕士生导师。
蔡吕培，中国矿业大学，硕士研究生。

应用可切换百叶窗系统的教室视觉舒适度优化研究

Research on Visual Comfort Optimization of Classroom Using Glazing System with Switchable Blinds

遇大兴，冯丽明，张祖玮
Yu Daxing, Feng Liming, Zhang Zuwei

摘要 在使用天然光照明的教室中，视看黑板时产生视觉舒适度欠佳的问题，主要由黑板面反射眩光引起的文字对比度下降导致。本文提出一种可切换百叶窗系统，旨在解决黑板亮度对比度下降导致的视觉不舒适问题，并改善窗视野清晰度。用双向散射分布函数（BSDF）描述提出的复杂开窗系统（CFS），并对优化后的室内采光性能的采光阈占比（sDA）以及黑板亮度对比度进行模拟，用“开放系数”评估视野清晰度。结论：使用可切换百叶窗系统可以在保持一定的天然光可用性的同时，显著提高黑板亮度对比度以及视野清晰度，为改善教室室内光环境提供一种可行方案。

关键词 天然采光；教室空间；视觉舒适度；百叶窗；光环境优化

Abstract In classrooms that use daylighting, poor visual comfort when looking at the blackboard is mainly caused by the decrease in text contrast caused by glare reflected on the blackboard surface. This paper proposes a switchable blinds system, aiming to solve the problem of visual discomfort of low contrast of the blackboard and to improve the window vision. The proposed CFS was described by BSDF, together with a simulation of the spatial Daylight Autonomy(sDA) and blackboard brightness contrast using the proposed system, while the visual field clarity was evaluated by the “openness factor”. Conclusion: The use of a switchable blinds system can significantly improve the brightness contrast of the blackboard and visual clarity while maintaining a certain daylight availability, which provides a feasible solution for improving the indoor light environment of the classroom.

Keywords Daylight; Classroom space; Visual comfort; Blinds; Light environment optimization

遇大兴，华南理工大学建筑学院，副教授。
冯丽明，华南理工大学建筑学院，硕士研究生。
张祖玮，华南理工大学建筑学院，硕士研究生。

夏热冬暖地区既有办公建筑外窗加装百叶低碳改造研究

Study on Low-carbon Renovation of Exterior Windows of Existing Office Buildings in Hot Summer and Warm Winter Areas

黄文捷，孟庆林
Huang Wenjie, Meng Qinglin

摘要 我国既有建筑存量极大，其中公共建筑能耗占总能耗比重大，亟需对其进行节能低碳改造。夏热冬暖地区既有办公建筑外窗热工性能不佳，多采用铝合金门窗搭配单层白玻的配置，导致该地区能耗水平较高。本文以能耗与视觉舒适度作为评价指标，通过 EnergyPlus 模拟比较了既有外窗加装百叶与整窗替换两种措施的节能效果与减碳效益。结果表明，外窗加装百叶与整窗替换两种措施的最大节能率分别为 25.13%、26.11%，两者差异不大。此外，外窗加装百叶的减碳效益反而优于整窗替换，具有更大的低碳改造潜力。

关键词 夏热冬暖地区；既有办公建筑；外窗低碳改造；能耗模拟

Abstract There is a huge stock of existing buildings in China, and the energy consumption of public buildings accounts for a large proportion of the total energy consumption, so it is urgent to carry out energy-saving and low-carbon transformation. The thermal performance of exterior windows of existing office buildings in hot summer and warm winter areas is not good, and aluminum alloy doors and windows are mostly equipped with single-layer white glass, which leads to high energy consumption in this area. Taking energy consumption and visual comfort as evaluation indexes, this paper compares the energy-saving effect and carbon-reducing benefit of adding louvers to existing external windows and replacing the whole window through EnergyPlus simulation. The results show that the maximum energy-saving rates of installing louvers on external windows and replacing whole windows are 25.13% and 26.11%, respectively, and there is little difference between them. In addition, the carbon reduction benefit of adding louvers to external windows is better than that of replacing the whole window, which has greater potential for low-carbon transformation.

Keywords Hot summer and warm winter areas; Existing office building; Low-carbon transformation of external windows; Energy consumption simulation

黄文捷，华南理工大学建筑学院，硕士研究生。
孟庆林，华南理工大学建筑学院，教授，亚热带建筑科学国家重点实验室副主任，中国建筑学会建筑物理分会理事长。

气凝胶玻璃采光性能及视觉舒适性实验研究

Experimental Study on the Daylighting Performance and Visual Comfort of Aerogel Glazing System

郑东梅，陈友明
Zheng Dongmei, Chen Youming

摘要 气凝胶玻璃具有很大的节能潜力，但对其采光性能和视觉舒适性的研究甚少。本文建立了气凝胶玻璃的全尺寸实验平台，在晴天和阴天测量房间水平照度，计算照度均匀度；通过对比测量和模拟的垂直眼照度，验证高动态范围（HDR）图像的准确性；采用 Evalglare 软件模拟三种眩光源检测方法（工作面法、阈值法、因子法）下的不舒适眩光可能性（DGP）和不舒适眩光指数（DGI）。结果表明，气凝胶玻璃房间在晴天和阴天均具有较高的照度均匀度。但气凝胶玻璃亮度过大，气凝胶玻璃房间在晴天可能存在不可忍受或干扰的眩光。

关键词 气凝胶玻璃；实验；采光；视觉舒适性；高动态范围图像

Abstract Granular aerogel glazing system (AGS) has great energy-saving potential, whereas the studies on the daylighting performance and the visual comfort of AGS are very limited. In this study, a full-scale experiment with an AGS window was conducted. The horizontal illuminance was monitored in overcast and sunny days. The illuminance uniformity was calculated. The high dynamic range (HDR) photograph technique was used and validated by comparing the measured and simulated vertical illuminance at eye level. The daylight glare probability (DGP) and daylight glare index (DGI) identified by three glare identification methods (task area method, threshold method, and factor method) were extracted from HDR images by Evalglare. The results showed that the illuminance uniformity in the AGS test room is higher in sunny and overcast days. It is also found that glare may occur in the AGS test room mainly due to the excessive luminance of the AGS window. The discomfort glare in the AGS test room may be intolerable or disturbing in sunny days.

Keywords Aerogel glazing system; Experiment; Daylighting; Visual comfort; High dynamic range photograph

郑东梅，湖南大学土木工程学院，博士研究生。
陈友明，湖南大学土木工程学院，高级教授。

基于太阳辐射分析的建筑外遮阳系数简化计算优化研究

Research on Optimization of Simplified Calculation to Building External Shading Coefficient Based on the Analysis of Solar Radiation

罗臣佑，于晓敏
Luo Chenyou, Yu Xiaomin

摘要 在工程应用中，建筑外遮阳系数一般由各气候分区下居建节能设计标准附录中的简化计算公式确定，或由现行国家节能规范（GB 55015—2021）给出的太阳辐射分析方法进行计算。本文以全国14个位于不同省级行政区的夏热冬冷A区城镇为分析对象，基于Grasshopper参数化设计平台和Radiance建筑光环境仿真工具推导并对比分析外遮阳特征值与外遮阳系数的拟合曲线及其方程。结果表明倒数函数方程相较二次函数方程对于简化计算而言更为准确，但建筑外遮阳系数的地域性相关因素仍需进一步探究。

关键词 太阳辐射；外遮阳系数；外遮阳特征值；拟合曲线

Abstract In engineering application, building external shading coefficient is generally determined by the simplified calculation formula in the Appendixes of the design standards for energy efficiency of residential buildings in different climate zones, or by the calculation method of solar radiation from the current national energy efficiency code (GB 55015—2021). This study takes 14 cities from different provincial administrative regions located in hot summer and cold winter zone (A) as research objects, deduces and compares the fit curves and equations of the projection factor and the external shading coefficient based on the parametric design platform named Grasshopper and building light environment simulation tool named Radiance. The results show that reciprocal function is more accurate than the quadratic function for simplified calculation to building external shading coefficient while further research is required on the regional related factors.

Keywords Solar radiation; External shading coefficient; Projection factor; Fit curve

罗臣佑，中国建筑西南设计研究院有限公司，工程师，设计师。
于晓敏，中国建筑西南设计研究院有限公司，工程师，设计师。

新型主动式光伏相变窗的光热电性能实测与节能研究

The Thermal Performance and Energy Saving of New Active Photovoltaic Window Integrated with Phase Change Material

龙其浩，李春莹*，唐海达
Long Qihao, Li Chunying, Tang Haida

摘要 本研究提出新型主动式光伏相变窗（PV-PCM window），可开闭的相变叶片不仅对该新型窗整体透光性影响小，还能按需利用相变材料的潜热，同时相变叶片通过相变蓄冷调节光伏玻璃温度优化其电性能。开展光伏相变窗与普通玻璃的光学、热学和电学性能对比试验，在深圳进行的实验结果为：相较于普通玻璃，室内光照强度最高可达2000 lx/m^2；光伏相变窗降温峰值达6.77℃；同时光伏相变窗的光伏发电效率比光伏窗高27.20%。发现光伏相变窗有助于保持良好室内热舒适并提高光伏玻璃性能。

关键词 光伏相变窗；建筑光伏一体化；实验研究；建筑节能

Abstract This study proposes a novel active photovoltaic phase change window (PV-PCM window) with operable phase change blades. The phase change blades not only have minimal impact on the overall transparency of the window but also enable the utilization of latent heat from the phase change material as per demand. Additionally, the phase change blades optimize the electrical performance of the photovoltaic glass by regulating its temperature through phase change cold storage. Comparative experiments were conducted to evaluate the optical, thermal, and electrical properties of the PV-PCM window in comparison to conventional glass. Experimental results conducted in Shenzhen reveal that compared to conventional glass, the indoor illuminance can reach a maximum of 2000 lx/m^2. The PV-PCM window exhibits a cooling peak of 6.77℃, and its photovoltaic power generation efficiency is 27.20% higher than that of the photovoltaic window. The results indicate that the PV-PCM window contributes to maintaining a comfortable indoor thermal environment while enhancing the performance of the photovoltaic glass.

Keywords PV-PCM window; Building integrated photovoltaic; Experiment; Building energy conservation

龙其浩，深圳大学建筑与城市规划学院，深圳市医养建筑重点实验室（筹建启动），硕士研究生。
李春莹，深圳大学建筑与城市规划学院，深圳市医养建筑重点实验室（筹建启动），副教授。
唐海达，深圳大学建筑与城市规划学院，深圳市医养建筑重点实验室（筹建启动），副教授，硕士生导师。

窗视野质量评估在建筑采光分析中的应用

The Application of Window View Quality Assessment in Daylighting Analysis for Buildings

赵永慧，边宇
Zhao Yonghui, Bian Yu

摘要 建筑天然光环境分析应包含三个主要方面：①天然光利用率；②视觉舒适度；③窗视野。良好窗视野对于房间使用者的身心均有益处，但目前窗视野尚缺少具体评价。本文系统介绍窗视野分析所涉及的内容及其指标和规定，并以 LEED v4.1 和 EN17037 建筑评价体系为例介绍，最后提取窗视野评价的重要特征及设计策略，提供将窗视野评价融入建筑设计过程的具体建议。

关键词 窗视野；视野内容；视野可见性；视野清晰度

Abstract The analysis of the daylight environment of a building should include three main aspects: ① daylight availability; ② visual comfort; ③ window view. A good window view is beneficial to both the mind and body of the room users, but there is no specific evaluation of window view. This paper systematically introduces the elements involved in window view analysis and its indicators and regulations. It also introduces the window view evaluation system using LEED v4.1 and EN17037 building evaluation systems as examples; finally, the important features of window view evaluation and design strategies are extracted to provide concrete suggestions for integrating window view evaluation into the architectural design process.

Keywords Window view; View content; View access; View clarity

赵永慧，华南理工大学建筑学院亚热带国家重点实验室，硕士研究生。
边宇，华南理工大学建筑学院亚热带国家重点实验室，副教授。

基于 Rhino + Grasshopper 的建筑幕墙光热聚焦模拟

Light and Heat Focusing Simulation of Building Curtain Wall Based on Rhino+Grasshopper

李越洋，高庆龙，陈姝亚，欧阳金龙
Li Yueyang, Gao Qinglong, Chen Shuya, OuYang Jinlong

摘要 我国有大量的高反射率建筑幕墙，镜面反射特性使得建高反射率建筑幕墙易发生光热聚焦的现象。如何建立建筑幕墙光热聚焦高效模拟模型，成为控制建筑幕墙光热聚焦现象的关键。本研究以 Rhino 软件为建模工具，以 grasshopper 为平台，使用 VBA 语言编写相关“电池”模块，建立了建筑幕墙光热聚焦动态模拟计算模型。它用以预测工程项目玻璃幕墙光热聚焦现象发生与否，光热聚焦产生时刻、位置以及主导玻璃位置参数。满足了模拟计算模型的高效性、轻量化、建筑专业友好化要求。最后结合某工程进行了实例验证。

关键词 建筑幕墙；高反射率；光热聚焦；Rhino-Grasshopper 平台；模拟模型

Abstract There are a large number of high-reflectivity building curtain walls in China, and the specular reflection characteristics make building high-reflectivity building curtain walls prone to light and heat focusing. How to establish an efficient simulation model of building curtain wall light and heat focusing has become the key to controlling the phenomenon of building curtain wall light and heat focusing. This study uses Rhino software as a modeling tool, grasshopper as a platform, and uses VBA language to write related "battery" modules, and establishes a dynamic simulation calculation model for building curtain wall light and heat focusing. It is used to determine whether the project glass will produce light and heat focus, when and where does light and heat focus appear, and which glasses play a major role in it. It meets the requirements of high efficiency, light weight, and professional friendliness of the simulation calculation model. Finally, an example is verified with a certain project.

Keywords Building curtain wall; High Reflectivity; Light and heat focusing; Rhino-Grasshopper platform; Simulation model

李越洋，四川大学，硕士研究生。
高庆龙，四川大学，教授。
陈姝亚，四川大学，硕士研究生。
欧阳金龙，四川大学，教授。

双层玻璃幕墙传热系数现场实测研究

On-site Measurement of the Heat Transfer Coefficient of a Double Skin Façade

刘人华，何国青 *，孟源
Liu Renhua, He Guoqing, Meng Yuan

刘人华，浙江大学建筑工程学院，硕士研究生。
何国青，浙江大学建筑工程学院，浙江大学平衡建筑研究中心，教授。
孟源，浙江大学建筑工程学院，硕士研究生。

摘要 本研究对杭州一间实际房间的双层玻璃幕墙进行了传热系数的测量和误差分析。选择室内外温度恒定的时间段，在保温性能良好的测试房间建立了稳态传热模型，用于计算通过双层玻璃幕墙的传热量。结果表明，双层玻璃幕墙的传热系数为 1.61 W/(m^2·K)。进一步分析得到，该值对测试房间与周围环境之间的温差较为敏感。误差分析显示，只要能够准确测量室内与环境之间的温差，这种现场测量方法对于测量双层玻璃幕墙的传热系数是可行的。

关键词 双层玻璃幕墙；传热系数；现场实验；误差分析

Abstract The heat transfer coefficient and error analysis of a double skin façade (DSF) in a real room in Hangzhou were presented. By selecting a period with constant indoor and outdoor temperatures, a steady-state heat transfer for the well-insulated test room was established to calculate of the heat transfer rate through the DSF. The result shows that the DSF has a heat transfer coefficient of 1.61 W/ (m^2·K). Further analysis shows that this value is sensitive to the temperature difference between the test room and the surrounding environments. Our error analysis shows that this field measurement method can be reliable for measuring the heat transfer coefficient of the DSF, as long as the temperature difference between the room and its surroundings is measured with some accuracy.

Keywords Double skin façade; Heat transfer coefficient; Field experiment; Error analysis

---

窗墙比和控制策略对电致变色窗户应用性能的影响

Effect of Window-to-wall and Control Strategy on Application Performance of Electrochromic Windows

武双对，孙弘历，黄志轩，田琳玥，林波荣
Wu Shuangdui, Sun Hongli, Huang Zhixuan, Tian Linyue, Lin Borong

武双对，清华大学建筑学院，博士研究生。
孙弘历，四川大学建筑与环境学院，副研究员。
黄志轩，田琳玥，四川大学建筑与环境学院，本科生。
林波荣，清华大学建筑学院，教授，副院长。

摘要 电致变色窗户具有可变的太阳辐射透过率，是建筑节能领域的研究热点。控制策略对电致变色窗户的应用性能有重要影响。本文以北京市 3 层办公建筑为例，通过 EnergyPlus 软件实现基于室外温度的控制策略，并以全年能耗、光热环境和光热舒适为评价指标，探究了窗墙比和控制策略对电致变色窗户应用性能的影响。研究结果表明，窗墙比和控制策略阈值的选取应寻求空调能耗和照明能耗的最佳平衡点，并满足舒适环境营造的需求。本文基于节能和舒适二者平衡的角度，为电致变色窗户的应用和性能优化提供了参考。

关键词 电致变色窗户；窗墙比；控制策略；建筑模拟；智能窗

Abstract Electrochromic windows with variable solar radiation transmittance are a hot research topic in the field of building energy efficiency. The control strategy has an important impact on the application performance of electrochromic windows. Taking a 3-story office building in Beijing as an example, this paper implements a control strategy based on outdoor temperature using EnergyPlus software and explores the effect of window-to-wall ratio and control strategy on the application performance of electrochromic windows, and using annual energy consumption, light and thermal environment and comfort as evaluation indexes. The results show that the WWR and the optimal control threshold should seek the best balance between air conditioning and lighting energy consumption, and meet the demand for comfortable environment built. Based on the balance between energy saving and comfort, this paper provides a reference for the application and performance optimization of electrochromic windows.

Keywords Electrochromic windows; Window-to-wall ratio; Control strategy; Building simulation; Smart windows

基于光学性能的3D打印建筑幕墙设计研究

Research on 3D-Printed Facade Based on Optical Performance

尹佳文，华好
Yin Jiawen，Hua Hao

摘要 现代建筑愈加重视立面的美观、采光和节能，其围护结构的材料也越来越丰富。本文以热熔塑料颗粒为原材料，研究使用大型 3D 打印技术制造的建筑幕墙的光学性能优势。3D 打印建筑幕墙在满足光学性能需求的情况下，结合针对性设计实现其物理学性能的优势，并赋予建筑表皮复杂造型与特殊纹理。使用 3D 打印技术能够在无模具、无支撑条件下，低成本、高效率地建造复杂几何曲面建筑幕墙，便于安装拆卸、重复利用，能很好地适配建筑师设计的非线性建筑。

关键词 3D 打印；建筑幕墙；热熔塑料；光学性能；打印路径规划

Abstract Modern buildings pay more and more attention to the beauty of facades, lighting and energy saving, and the materials of their envelope structures are also becoming more and more abundant. In this study, the optical performance advantages of building curtain walls manufactured using large-scale 3D printing technology using hot-melt plastic particles as raw materials. The 3D printed building curtain wall is designed to meet the requirements of optical performance, combined with targeted design to realize the advantages of its physical performance, and endow the building skin with complex shapes and special textures. The use of 3D printing technology can build complex geometric curved building curtain walls at low cost and high efficiency without molds and supports, which is easy to install, disassemble and reuse, and can well adapt to nonlinear buildings designed by architects.

Keywords 3D printing; Building curtain wall; Hot melt plastic; Optical properties; Toolpath planning

尹佳文，东南大学建筑学院，硕士研究生。
华好，东南大学建筑学院，副教授。

气候变化对中国绿色建筑围护结构标准的影响

The Impact of Climate Change on Green Building Envelope Standards in China

Akram Ahmed Noman Alabsi，杜峰，刘辉，蔡堂生，徐雅倩
Akram Ahmed Noman Alabsi, Du Feng, Liu Hui, Cai Tangsheng, Xu Yaqian

摘要 本文研究了气候变化对绿色参考建筑能源性能的影响。在评估所选研究区域的预期室外温度趋势后，通过建筑围护结构的设计和实施对能耗产生影响进行分析。在六个地点进行能量动力学模拟，以覆盖中央气候区域的气候波动。使用反映气候变化加剧的两种情景来检验 2030 年和 2060 年的未来气候情景。研究结果显示，在 2030 年，部分南方和东部地区的冬季供暖需求将显著减少，可能超过 30%。同时，部分中部和北方地区的夏季制冷需求将显著增加，超过 50%。随着冬季和夏季能源需求的变化趋势，有针对性的政策在减少建筑能耗方面发挥关键作用，对于中国实现“双碳”目标至关重要。

关键词 围护结构；气候变化；建筑减碳；建筑节能

Abstract This study examines the impact of climate change on the energy performance of green reference buildings and focuses on the building envelope as a key factor influencing energy consumption. By evaluating projected trends in outdoor temperatures for selected regions, we conducted energy dynamics simulations in six locations representing central climatic regions in China. Two climate change tracks were considered for future climate scenarios in 2030 and 2060. Our findings indicate including a notable reduction in winter heating demand by over 30% in certain southern and eastern areas by 2030. Conversely, cooling demand will increase significantly in the summer, surpassing 50% in central and northern China regions. As total energy requirements exhibit evolving trends in winter and summer, targeted policies to decrease building consumption play a crucial role in China's attainment of the carbon peaking and carbon neutrality goals.

Keywords Building envelope; Climate change; Building stock decarbonization; Energy efficiency

Akram Ahmed Noman Alabsi，福建理工大学建筑与城乡规划学院，副教授，硕士生导师。
杜峰，福建理工大学建筑与城乡规划学院，副教授，建筑学专业负责人。
刘辉，福建理工大学建筑与城乡规划学院，讲师。
蔡堂生，徐雅倩，福建理工大学建筑与城乡规划学院，本科生。

“双碳”背景下闽东南沿海城市通风光伏屋顶设计策略研究

Research on the Design Strategy of Ventilated Photovoltaic Roof in Coastal Cities in Southeast Fujian in the context of Carbon Peaking and Carbon Neutrality Goals

杜峰，蔡晓静，郭培凌
Du Feng, Cai Xiaojing, Guo Peiling

摘要 “双碳”目标背景下，光伏发电作为减碳有效手段快速发展。传统通风屋顶因达不到理想效益逐渐被淘汰。本文以福州市为例通过考虑风向指标和太阳高度角，研究一种适用于东南沿海城市的通风光伏屋顶。采用CiteSpace数据分析和PHOENICS软件数据模拟检验研究假设，得出以下结果：新型可调节通风光伏屋顶能最大程度利用太阳辐射实现产能效益，并且根据季节调整安置角度能够有效改善建筑室内热环境，对“双碳”目标下未来城市建筑更新有良好的推进作用。

关键词 通风光伏屋顶；“双碳”目标；东南沿海；城市更新；设计策略

Abstract In the context of carbon peaking and carbon neutrality goals, photovoltaic power generation is developing rapidly as an effective means of carbon reduction. The traditional ventilation roof is gradually eliminated because it cannot reach the ideal benefit. Taking Fuzhou area as an example, this paper studies a ventilation photovoltaic roof suitable for southeast coastal cities by considering wind direction index and solar height Angle. Using CiteSpace analysis and PHOENICS software data simulation to test the research hypothesis, the following results were obtained: The new type of adjustable ventilation photovoltaic roof can maximize the use of solar radiation to achieve productivity benefits, and can adjust according to season, effectively improve indoor thermal environment of the building, and has a good role in promoting the future urban building renewal under the goals of carbon peaking and carbon neutrality.

Keywords Ventilated Photovoltaic roof; Ccarbon peaking and carbon neutrality goals; Southeast Coast; Urban Renewal; Design Strategy

杜峰，福建理工大学建筑与城乡规划学院，副教授。
蔡晓静，郭培凌，福建理工大学建筑与城乡规划学院，本科生。

基于光热协同控制的动态遮阳系统分析研究

Analysis and Research on Dynamic Shading System Based on Synergistic Control of Light and Thermal

高亚萍，高岩 *，范雨佳
Gao Yaping, Gao Yan, Fan Yujia

摘要 适应环境变化的遮阳系统是助力我国实现“双碳”目标的重要发展方向。本研究对北京市太阳辐射及日照情况进行分析，基于光线追踪算法，采用 Matlab 和 DeST 进行仿真模拟。本研究在考虑室内温度和照度协同评价的同时，对不同条件下光、热环境进行分析。结果表明动态遮阳系统具有显著优越性，室外温度达到峰值时，温差显著增大（1.9~4.6℃），起到了能耗峰值降低的作用，与此同时，视觉舒适度提高 28%~37%，热舒适度提高 27%~43%。本研究改善了室内光、热环境，挖掘了建筑围护结构的节能潜力。

关键词 围护结构；遮阳；室内温度；节能

Abstract The shading system is an important development direction to help China achieve carbon peaking and carbon neutrality goals. The study analyzes the solar radiation and sunshine in Beijing, and calculates dynamic shading coefficients based on ray-tracing algorithm and simulation with Matlab and DeST. Considering the synergistic evaluation of indoor temperature and illumination, the simulation analysis of light and heat environment and building energy consumption under different conditions is carried out. The results show that the dynamic shading system has significant superiority, and the temperature difference increases significantly (1.9~4.6°C) when the outdoor temperature reaches its peak, which plays a role in the peak energy consumption reduction. At the same time, visual comfort is improved by 28%~37% and thermal comfort is improved by 27%~43%. This study tapped the energy-saving potential of the building envelope for improving the indoor light and thermal environment.

Keywords Envelope structure; Shading; Indoor temperature; Energy saving

高亚萍，北京建筑大学，博士研究生。
高岩，北京建筑大学，教授。
范雨佳，北京建筑大学，博士研究生。

第 6 篇

城市微气候

面向城市广场微气候改善的水体模拟设计研究

Study on Water Simulation Design for Microclimate Improvement of Urban Square

周雅婷，赖叶红，陈睿智 *
Zhou Yating, Lai Yehong, Chen Ruizhi

周雅婷，湖南大学建筑与规划学院丘陵城市规划研究中心，硕士研究生。
赖叶红，京都大学地球环境学堂人间环境设计论分野，硕士研究生。
陈睿智，西南交通大学建筑学院，副教授。

摘要 城市热岛效应日益严重，水体作为重要的景观要素之一，对城市微气候的改善起着重要作用。本研究以为民广场为研究对象建立实际的三维数值模拟模型，使用 ENVI-met 三维非静力气候模拟软件，对水体微气候特征进行模拟分析，结果表明：城市中的水体对其周边微气候有着明显的调节作用；水体降温作用与其斑块面积比呈对数函数关系，考虑到经济效应，最适斑块面积比不宜过大；等面积水体下，分散度与降温增湿作用呈负相关，与辐射面积呈正相关。

关键词 微气候；城市广场；ENVI-met；水体布局要素

Abstract The Heat Island effect is increasingly serious. As one of the important landscape elements, water plays an important role in improving the urban microclimate. In this study, Weimin Square was taken as the research object to establish an actual 3d numerical simulation model. ENVI-met 3d non-static climate simulation software was used to simulate and analyze the characteristics of water microclimate. The results indicate that: Water has an obvious regulating effect on the surrounding microclimate; The relationship between the water cooling effect and the plaque area ratio is present as a logarithmic function. Considering the economic effect, the optimal plaque area ratio should not be too large; Under the same area of water, the dispersion degree is negatively correlated with the cooling and humidifying effect, but positively correlated with radiation area.

Keywords Microclimate; Residential square; ENVI-met; Water landscape layout

---

室外空间微气候逆向设计——以天津市某幼儿园为例

Inverse Design the Microclimate of Outdoor Space—Kindergarten in Tianjin as an Example

孙睿，刘炜，赖达祎 *
Sun Rui, Liu Wei, Lai Dayi

孙睿，上海交通大学，硕士研究生。
刘炜，天津大学环境科学与工程学院，教授；瑞典皇家理工学院土木与建筑工程系，副教授。
赖达祎，上海交通大学设计学院，上海交通大学未来技术学院，副教授。

摘要 舒适的室外微气候对于创造优质的户外空间至关重要。本文提出优化建筑形状、朝向和位置的逆向设计框架，以缓解目标场地室外冷热应力，并综合考虑周边环境对目标场地的影响和全年气候变化。随后，本文将该框架应用于天津气候条件下的幼儿园室外空间设计。结果表明该框架优化改善了目标场地的室外热舒适。本文丰富了建筑尺度室外热舒适的逆向设计研究，并为营造舒适的城市室外空间提供了理论指导。

关键词 逆向设计；室外热舒适；遗传算法；计算流体力学

Abstract Thermally comfortable microclimate is essential for creating high-quality outdoor spaces. This study proposes an optimization framework to determine the building shape, orientation, and location that reduces the overall thermal stress in the target outdoor space. The framework considers both the effect of surrounding buildings and seasonal changes. Then, this investigation applied the proposed optimization framework to design the outdoor space for a kindergarten in the climate of Tianjin. The results showed that optimization improved the outdoor thermal comfort conditions. This study supplements the inverse design of outdoor thermal comfort at the building scale and provides suggestions to create comfortable urban outdoor spaces.

Keywords Inverse design; Outdoor thermal comfort; Genetic algorithm; CFD

城市微尺度植被物理过程测试及模型研究进展

A Review of Urban Micro-scale Vegetation Physical Process Testing and Modeling

燕俊儒，赵立华 *，张磊
Yan Junru, Zhao Lihua, Zhang Lei

摘要 植被是影响城市微气候的关键地物类型，由于复杂的物理和生理过程及环境扰动，植被与环境的相互作用机制仍然难以量化。目前主流数值模拟工具将中尺度气象模型部分子模型直接应用于微尺度气候模拟，但由于不同尺度之间规律及其现象总呈现不可忽视的差异，这可能造成模拟结果的误差从而误导设计。本文梳理气候尺度界定的演变，总结近年微尺度植被物理过程测试和植被模型开发，提出跨尺度植被模型应用存在的问题和局限，为后续微尺度植被模型研究和开发提供支持，以期助力城市微气候的循证和可持续设计。

关键词 城市微气候；植被模型；尺度推绎；循证设计；可持续设计

Abstract Vegetation is a key feature type affecting urban microclimate, and the mechanisms of vegetation-environment interactions remain challenging to quantify due to complex physical and physiological processes and environmental perturbations. At present, numerical simulation tools apply some submodels of mesoscale meteorological models directly to micro-scale climate simulations, but since the laws and their phenomena always show non-negligible differences between different scales, this may cause errors in simulation results and thus mislead the design. This paper compares the evolution of climate scale definition, summarizes the physical process testing and vegetation model development of micro-scale, and presents the problems and limitations of cross-scale vegetation model application to support the subsequent research and development of micro-scale vegetation models, in order to support the evidence-based and sustainable design of urban micro-climate.

Keywords Urban microclimate; Plant modeling; Scaling; Evidence-based design; Sustainable design

燕俊儒，华南理工大学，博士研究生。
赵立华，华南理工大学，教授。
张磊，华南理工大学，副研究员。

不同降雨事件中风驱雨测量误差研究

Measurement Errors of Wind-driven Rain in Different Rain Events

路标，李蕴洁，张会波，高亚锋，冯驰 *
Lu Biao, Li Yunjie, Zhang Huibo, Gao Yafeng, Feng Chi*

摘要 现场实测是确定建筑立面的风驱雨量及其分布规律最直接的方法。在不同降雨事件中，风驱雨的测量误差存在明显差别。本文研究风驱雨的测量误差来源，通过现场实测分析了不同降雨事件中的风驱雨测量误差。结果表明，在风速和风向稳定且降雨连续的降雨事件中风驱雨测量误差较小；而降雨事件中间断期越多，产生的绝对误差越大。对于总降雨量小的降雨事件，风驱雨测量误差可达 100%。盛行风向下建筑立面风驱雨测量的相对误差小于其他立面。

关键词 风驱雨；现场实测；风速；风向；降雨量

Abstract Filed measurement is the most direct method to determine the wind-driven rain (WDR) amount and its distribution on building facades. The measurement errors of WDR vary significantly in different rain events. This paper studies the sources of WDR measurement errors and analyses these errors in different rain events through field measurements. The results show that WDR measurement errors are small in continuous rain events with stable wind velocity and direction. On the contrary, rain events with many intermittent periods lead to large errors. For events with small rain amount, the WDR measurement errors can be up to 100%. The relative error of WDR measurements for the building facade in the prevailing wind direction is smaller than for other directions.

Keywords Wind-driven rain; Field measurement; Wind velocity; Wind direction; Rain amount

路标，重庆大学建筑城规学院，硕士研究生。
李蕴洁，重庆大学建筑城规学院，硕士研究生。
张会波，上海交通大学设计学院，副教授。
高亚锋，重庆大学土木工程学院，教授。
冯驰，重庆大学建筑城规学院，教授，建筑技术科学系副系主任。

严寒地区冬季城市开放空间形态要素与气温相关性研究

Study on the Correlation between Urban Open Space Morphology and Air Temperature in Winter in Severe Cold Regions

单忱璐，秦淦
Shan Chenlu, Qin Gan

单忱璐，哈尔滨工业大学，硕士研究生。
秦淦，哈尔滨工业大学，硕士研究生。

摘要 严寒地区冬季气温低且持续期长，导致城市开放空间活力不足。开放空间内微气候受城市形态影响显著，且二者关系具有高度动态性与复杂性。本文以哈尔滨为例，设置 20 个测点进行实测研究，发现冬季各测点气温与气象站参考气温差值在 -1.41 ~ 4.48℃。基于随机森林建立严寒地区冬季的开放空间形态要素与气温关系模型，并进行依赖关系分析发现，各城市形态要素与气温呈明显的非线性关系，其中距水体距离对气温影响最为显著，天空开阔度次之，绿容率对气温影响最为有限。

关键词 城市开放空间；严寒地区；微气候；随机森林；气温

Abstract In severe cold regions, the low temperature and long duration of winter result in insufficient vitality of urban open spaces. The microclimate in open spaces is significantly affected by urban morphology, and the relationship between them is highly dynamic and complex. Taking Harbin as an example, this paper sets up 20 measuring points for field research and finds that the difference between the air temperature of each measuring point and the reference temperature of the meteorological station ranges from -1.41℃ to 4.48℃ in winter. Based on the Random Forest, a relationship model between open space morphology and air temperature is established and a dependency analysis is conducted. The results show that there is an obvious nonlinear relationship between urban morphology and air temperature in winter. Among them, Distance to Waterbody (DW) has the most significant impact on temperature, followed by Sky View Factor (SVF), and Green Plot Ratio (GnPR) has the least impact.

Keywords Urban open space morphology; Severe cold region; Microclimate; Random Forest; Air temperature

---

基于大数据分析的露营行为的绿地需求研究

Research on Public Green Space Demand Based on Big Data Analysis of Camping Behavior

李丽 *，周婷婷
Li Li, Zhou Tingting

李丽，广州大学建筑与城市规划学院，副教授，硕士生导师。
周婷婷，广州大学建筑与城市规划学院，硕士研究生。

摘要 近年来，城市露营已成为新兴的户外活动形式，既有城市绿地不能满足人群日益增长的露营行为需求。本文采用大数据分析方法，以 2019 年疫情起至 2023 年初广州微博用户露营相关的 5640 条言论作为数据样本库，对露营行为的绿地需求进行研究。本文主要分析了广州城市露营情绪变化、露营行为新词与关键词以及露营微博话题主题内容。通过以上研究，明确人群的绿地需求主要有亲自然和美景度、户外活动和互动交流、空间可达性、服务设施和生态科技四个方面，并针对需求提出了对应绿地建设的建议。

关键词 社交媒体；大数据；文本挖掘；露营行为；绿地需求；风景园林

Abstract In recent years, urban camping has become a new form of outdoor activities, but the existing urban green space cannot meet the growing needs of people' s camping behavior. In this paper, big data analysis method was used to study the green space demand of camping behavior by taking 5640 camping comments of Weibo users in Guangzhou from 2019 to the beginning of 2023 as a data sample database. It mainly analyzes the change in urban camping mood, the new words and keywords of camping behavior and the topic content of the camping microblog. Based on the above research, it is clear that people' s green space needs mainly include four aspects: pro-nature and beauty, outdoor activities and interaction, space accessibility, service facilities and ecological science and technology, and the corresponding green space construction suggestions are put forward according to the needs.

Keywords Social media; Big Data; Text mining; Camping behavior; Green space demand; Landscape architecture

不同高度居住建筑群的风、热环境外场实验初探

Scaled Outdoor Experimental Studies of Wind and Thermal Environments in Residential Buildings with Different Heights

董志昀，尤伟 *，陆孟荣，杭建

Dong Zhiyun, You Wei, Lu Mengrong, Hang Jian

董志昀，南京大学建筑与城市规划学院，硕士研究生。
尤伟，南京工业大学建筑学院，副教授。
陆孟荣，荷兰格罗宁根大学材料与工程学院，博士研究生。
杭建，中山大学大气科学学院，教授，副院长。

摘要 本文探讨不同形态类型的居住建筑群在风、热耦合作用下的室外通风和热环境。研究以高层（等高）以及多层、高层组合（非等高）住宅两种典型居住区作为研究对象，利用外场缩尺实验探寻了建筑高度变化对冬季室外风、热环境的影响。初步研究表明：在相同容积率、建筑密度的情况下，建筑高度变化是影响居住风热环境条件的一个重要因素。对于非等高建筑群，高度较低建筑的北墙面平均温度较等高建筑群相同位置墙面平均温度高，其温差可达 5.2℃。

关键词 居住建筑群；高度组合变化；风热耦合条件；外场缩尺实验；通风条件、热环境

Abstract This study discusses the outdoor ventilation conditions and thermal environment of different types of residential buildings under wind and thermal coupling. The research takes two typical residential areas of high-rise buildings （same height） and multi-story and high-rise composite buildings （not the same height） as the research objects and explores the impact of building height changes on the outdoor wind and thermal environment in winter through the scaled outdoor experiments. The preliminary study shows that under the same floor area ratio and building density condition, the building height change is an important factor affecting the residential wind-heat environmental conditions. Compared with the same height building group, the average temperature of the north wall of the lower buildings in the height combination building group is higher, and the temperature difference can reach 5.2℃ .

Keywords Residential buildings; Height change; Wind and heat coupling conditions; Scaled outdoor experiment; Ventilation conditions and thermal environment

---

基于局地气候分区的成都街道空间热舒适研究

Research on Thermal Comfort of Chengdu Streets Based on Local Climatic Zones

陈岚，郝又佳，欧阳金龙

Chen Lan, Hao Youjia, Ouyang Jinlong

陈岚，四川大学建筑与环境学院，副教授，硕士生导师。
郝又佳，四川大学建筑与环境学院，硕士研究生。
欧阳金龙，四川大学建筑与环境学院，教授，硕士生导师。

摘要 本文基于局地气候分区理论并结合林荫空间对街道的限定，在成都市选出 6 条符合相应指标的街道，在夏冬两季进行研究实验，结果表明：①在同时空下，不同 LCZ 分区下的街道微气候数值有差异；同类型街道在不同季节的微气候大小关系不同。②在冬季，紧凑低层型和开敞中层型是相对最为舒适的两类街道；在夏季，紧凑高层型和开敞高层型是最为舒适的街道类型。③不论冬夏，街道的使用群体对微气候热舒适均随热感觉增加而升高，在热舒适到达最值后，均会随热感觉值升高而降低。④街道空间形态会在一定程度上影响人的热偏好，进而改变人对热环境的心理预期与适应能力。

关键词 局地气候区；街道空间；微气候；热舒适；生理等效温度

Abstract Based on the theory of Local Climate Zone and the restriction of tree-lined space on streets, this paper selects six streets in Chengdu, and conducts research experiments in summer and winter, the results show that: ① Under the same space-time condition, the microclimate values of different LCZ districts are different; the microclimate of the same type of street varies in different seasons. ② In winter, the compact low-rise and the open middle-rise are the two most comfortable streets. In summer, compact high-rise and open high-rise are the most comfortable street types. ③ In both summer and winter, the thermal comfort of street users increased with the increase of thermal sensation, and decreased with the increase of thermal sensation when the thermal comfort reached its maximum value. ④ Street space will affect people' s thermal preference to a certain extent, and then change people' s psychological anticipation and adaptability to the thermal environment.

Keywords Local Climate Zone; Street space; Micro-climate; Thermal comfort; PET

基于 CiteSpace 的起伏地形下太阳辐射估算研究进展

Progress in the Study of Solar Radiation Under Complex Terrain Based on CiteSpace

韩冰，牛兵兵，曹其梦，杨柳，刘衍

Han Bing, Niu Bingbing, Cao Qimeng, Yang Liu, Liu Yan

——

韩冰，牛兵兵，西安建筑科技大学建筑学院，博士研究生。

曹其梦，西安建筑科技大学交叉创新研究院，助理教授。

杨柳，西安建筑科技大学建筑学院，教授。

刘衍，西安建筑科技大学建筑学院，教授。

摘要 研究对 1990 年以来起伏地形下太阳辐射估算研究刊载论文进行多方位统计整理，运用 CiteSpace 软件和知网文献检索功能探析近年来国内外相关研究的历程与发展趋势，结果表明：该领域大致经历了三个时期的发展，目前的热点主要聚焦地理信息数据的智能化表达以及卫星遥感数据的时空运用等。在太阳辐射分布式模型方面，“高分辨率、小气候、时空分布”等高频词汇表明，针对不同地表的太阳辐射对局地气候精细化影响下的研究正在逐步提升。本研究可为不同地形条件下太阳能资源的准确评估和利用等关键技术的提升提供相关参考。

关键词 起伏地形；太阳辐射；分布式模型；进展与展望

Abstract The research has compiled multi-faceted statistics on papers published since 1990 on solar radiation distribution models in complex terrain and explored the history and development trend of related research at home and abroad using CiteSpace software and CNKI literature search function. The current hotspots focus on the intelligent representation of geographic information data and the spatial and temporal application of satellite remote sensing data. In terms of solar radiation distribution models, "high resolution, microclimate, and spatial and temporal distribution" indicate that the refinement of the study of solar radiation on different surfaces is gradually increasing under the influence of local climate. This research can provide important reference values for the improvement of key technologies for the accurate assessment and utilization of solar energy resources under different terrain conditions in China.

Keywords Complex terrain; Solar radiation; Distributed models; Overview and prospect

---

沈阳中街步行街冬季热环境实测及模拟分析研究

Field Experiment and Simulation Analysis of Thermal Environment in Shenyang Centre Street in Winter

席天宇，曹恩嘉，王明，李瑾，张幸宇，芦乐平

Xi Tianyu, Cao Enjia, Wang Ming, Li Jin, Zhang Xingyu, Lu Leping

——

席天宇，东北大学，教授。

曹恩嘉，王明，李瑾，张幸宇，芦乐平，东北大学，硕士研究生。

摘要 室外商业步行街人流动线多数集中在室外休闲区域及步行交通区域，室外空间热环境成为影响商业步行街人群热舒适感受的关键因素。本文对沈阳市中街步行街的丁字路口、十字路口、层峡中心的空气温度、相对湿度、风速、黑球温度进行实测分析，并通过 ENVI-met 软件进行模拟与数据对比，实测和模拟结果对严寒地区步行街热环境及热舒适性相关研究具有参考价值。

关键词 热环境；ENVI-met；中街步行街；严寒地区

Abstract The flow lines of outdoor commercial pedestrian streets are mostly concentrated on outdoor leisure areas and pedestrian traffic areas, so the outdoor thermal environment becomes one of the key factors affecting people' s thermal comfort in a commercial pedestrian street. This paper measured and analyzed the air temperature, relative humidity, wind speed and black bulb temperature in the center of the T-intersection, intersection and street canyon of Shenyang Centre Street. Additionally, this paper simulated with ENVI-met and compares the physical factors among those different environment points. The measured and simulated results have reference values for the research on the thermal environment and thermal comfort of walking streets in cold regions.

Keywords Thermal environment; ENVI-met; Shenyang Centre Street; Severe cold regions

闽中地区幼儿园儿童冬季室内外活动热感觉差异性研究

A Study on the Differences in Thermal Sensation Between Indoor and Outdoor Activities in Winter Among Kindergarten Children in the Central Fujian Region

张子钦，冉茂宇
Zhang Ziqin, Ran Maoyu

摘要 本文选取闽中地区两所幼儿园作为研究对象，实测和问卷调查了幼儿冬季在室内外活动的热环境及热感觉。结论表明：①冬季，室内操作温度以及室内平均辐射温度是影响儿童室内热感觉差异的关键因素；室外平均辐射温度是影响儿童室外热感觉差异的关键因素。②儿童在室内外不同的活动水平也是造成其热感觉差异的原因，在幼儿代谢率较高的情况下也有较大的PMV值。③在两所幼儿园的比较中，解建幼儿园在热舒适性方面更加适合幼儿。

关键词 闽中地区；幼儿园儿童；室内外活动；热感觉

Abstract This paper selects two kindergartens in central Fujian and conducts actual measurements and questionnaire surveys on the thermal environment and sensation of children's indoor and outdoor activities in winter. The conclusion indicates that: ① Indoor operating temperature and average indoor radiation temperature are key factors affecting differences in indoor thermal sensation in winter; The average outdoor radiation temperature is a key factor affecting the differences in children's outdoor thermal sensation. ② The different levels of indoor and outdoor activity among children are also the reason for their differences in thermal sensation, and even when their metabolic rate is high, there is a higher PMV value. ③ In the comparison between the two kindergartens, Jiejian Kindergarten is more suitable for children in terms of thermal comfort.

Keywords Central Fujian region; Kindergarten children; Indoor and outdoor activities; Thermal sensation

张子钦，华侨大学，硕士研究生。
冉茂宇，华侨大学，博士生导师。

基于URANS的理想街区气体污染物传播动态过程模拟

URANS-Based Simulation of Dynamic Processes of Gas Pollutant Propagation in an Ideal Block

胡一帆，韩梦涛
Hu Yifan, Han Mengtao

摘要 城市污染物扩散是一个复杂的动态过程，经常需要了解污染物从产生到目标地的时空扩散规律，但目前城市污染物扩散CFD模拟主要为稳态模拟，难以针对污染物的时间发展进行研究。本文基于URANS模拟理想街区中污染物的动态扩散过程并与风洞试验进行比较，研究了污染物浓度持续释放及释放结束衰减的动态传播过程。结果表明，模拟与实验基本一致，非稳态污染模拟对整体扩散过程的模拟精度较好。本模拟使用的方法对污染物扩散的动态过程模拟有较好的适应性，可为污染事故传播过程、人口受灾暴露风险等研究提供较好的方法与数据基础。

关键词 Unsteady RANS；污染物扩散；城市街区；污染物浓度时序列

Abstract The diffusion of pollutants in urban areas is a complex dynamic process that often requires understanding the spatiotemporal diffusion patterns of pollutants from their sources to their target locations. This paper simulates the dynamic diffusion process of pollutants in an ideal block based on URANS and compares it with wind tunnel experiments to study the dynamic propagation process of continuous release and end-of-release decay of pollutant concentration. The results showed good agreement between the simulation and experimental data, with a high level of accuracy in URANS pollutant simulation. The method used in this study shows good adaptability to simulate the dynamic process of pollutant diffusion and provides a useful basis for research on the spread of pollution incidents and the exposure risk to affected populations.

Keywords Unsteady RANS; Pollutant diffusion; Urban block; Time history of pollutant concentration

胡一帆，华中科技大学建筑与城市规划学院，硕士研究生。
韩梦涛，华中科技大学建筑与城市规划学院，副研究员。

风环境对建筑能耗的影响机理与理论模型综述

A Review of Mechanisms and Theoretical Models for the Impact of Wind Environment on Building Energy Consumption

介鹏宇，石邢
Jie Pengyu, Shi Xing

介鹏宇，同济大学建筑与城市规划学院，博士研究生。
石邢，同济大学建筑与城市规划学院，高密度人居环境生态与节能教育部重点实验室，教授，副院长 / 主任。

摘要 城市气候环境和建筑能耗存在密不可分的耦合关系，风环境与城市热湿环境一样，其作为关键的微气候要素对建筑能耗具有显著的影响，有必要系统研究总结其影响关系。本文从单体建筑能耗层面出发，总结研究风环境影响建筑能耗的基本机理和理论模型，对建筑能耗模型中风环境的理论模型发展进行比较分析，对该领域现有研究和未来方向提出总结与思考。这些总结和研究能够为城市能源政策的制定和城市节能提供相应参考，有利于节能低碳城市的理论构建与实践研究。

关键词 风环境；城市微气候；建筑能耗模拟；影响机制

Abstract There exists an inextricable coupling relationship between the urban climate environment and building energy consumption. Wind environment, like urban heat and humidity environment, as a key microclimate element, has a significant impact on building energy consumption, and it is necessary to systematically study and summarise its impact relationship. In this paper, we summarise the basic mechanism and theoretical model of the wind environment affecting building energy consumption from the building scale, make a comparative analysis of the theoretical model development of the wind environment in the building energy consumption model. This paper provides a summary and reflections on existing research and future directions in the field. These summaries and research can provide corresponding references for the formulation of urban energy policies and urban energy conservation, which is conducive to the theoretical construction and practical research of energy-saving and low-carbon cities.

Keywords Wind environment; Urban microclimate; Building Energy Modeling; Impact mechanisms

---

半岛山地城市空间形态对热环境的影响及缓解策略

Effects of Urban Form on the Thermal Environment and Mitigation Strategies in the Peninsula Mountainous City

赵君，郭飞 *，张弘驰，董菁，郭若男
Zhao Jun, Guo Fei, Zhang Hongchi, Dong Jing, Guo Ruonan

赵君，大连理工大学建筑与艺术学院，博士研究生。
郭飞，大连理工大学建筑与艺术学院，教授，副院长。
张弘驰，大连理工大学建筑与艺术学院，讲师。
董菁，大连理工大学建筑与艺术学院，助理研究员。
郭若男，大连理工大学建筑与艺术学院，博士研究生。

摘要 在全球气候变化的背景下，半岛山地城市由于复杂的地理环境更容易受到不良天气的冲击。基于此，本文以半岛山地城市大连为例，探讨了热浪天气下特定城市空间因子对地表温度（LST）的影响。城市空间因子除土地覆盖和三维城市形态参数外，还计算了研究地块距最近海岸距离（DFC）和距山地距离（DFM）。结果如下：① 100 m 网格更适用于本研究；② DFC 和 DFM 对 LST 的影响距离分别可达到 5053 m 和 4503 m；③通过量化分析发现建筑密度、平均建筑高度、DFC 和 DFM 对 LST 有明显影响，空间回归模型也被证明可以评估规划策略对热环境的影响。

关键词 半岛山地城市；城市空间因子；缓解策略；地表温度

Abstract Under the context of global climate change, peninsular mountainous cities are more vulnerable to adverse weather impacts due to their complex geographical environment. Based on this, this study explores the influence of specific urban spatial factors on land surface temperature (LST) under heat wave weather, using the peninsular mountain city of Dalian as an example. In addition to the land cover and 3D urban morphology parameters, the distance to the nearest coast (DFC) and distance to the mountains (DFM) were calculated for the urban spatial factors. The results are as follows: ① a 100m grid is more suitable for this study; ② linear regression was used to explore the degree of influence of DFC and DFM on land surface temperature; ③ Through quantitative analysis, it was found that building density, average building height, DFC and DFM have a significant impact on LST, and spatial regression models have also been proven to evaluate the impact of planning strategies on the thermal environment.

Keywords Peninsula mountainous city; Urban spatial factors; Mitigation measures; Land surface temperature

基于流动观测的大连市春季热环境特征与热岛强度分析

Analysis of Thermal Environment Characteristics and Heat Island Intensity in Dalian Based on Mobile Observation

张弘驰，白金，赵君，郭飞 *
Zhang Hongchi, Bai Jin, Zhao Jun, Guo Fei

摘要 在全球变暖与快速城市化影响下，城市热岛对公共安全产生巨大影响。基于局地气候分区（LCZ）的方案已被证实能够评估城市热岛强度，但对复杂下垫面的异质性评价仍有待研究。本文采用流动观测法于春季昼间和夜间测试大连市南部海岸沿线区域观测线路上各点的空气温度值及热岛强度。分析结果显示：建成区 LCZ 全天 UHI 变化幅度明显高于自然型 LCZ，且紧凑类型通常比开放类型有更强的热岛强度；城市中水体对其周边的小气候具有明显的调节作用；由于地理位置、人为排热、邻近效应等因素的影响，局部区域同类 LCZ 之间会出现热差异的现象。

关键词 局地气候分区；移动观测；城市热岛；滨海地区

Abstract Under the influence of global warming and rapid urbanization, urban heat island has a huge impact on public safety. Schemes based on local climate zoning have been shown to assess urban heat island intensity, but heterogeneous evaluation of complex substrates remains to be studied. In this paper, the air temperature values and heat island intensity of each points along the observation line in the southern coast of Dalian City were measured during daytime and nighttime in spring using the mobile observation method. The results of the analysis show that: the magnitude of UHI variation throughout the day in LCZs in built-up areas is significantly higher than that in natural types of LCZs, and compact types usually have stronger heat island intensity than open types; the water body in the city has a significant moderating effect on the surrounding microclimate; due to factors such as geographic location, anthropogenic heat dissipation, and proximity effects, there will be thermal differences among LCZs of the same species in local areas.

Keywords Local climate zone; Mobile observation; Urban heat island; Coastal area

张弘驰，大连理工大学建筑与艺术学院，讲师。
白金，大连理工大学建筑与艺术学院，硕士研究生。
赵君，大连理工大学建筑与艺术学院，博士研究生。
郭飞，大连理工大学建筑与艺术学院，教授，副院长。

湿热城市高温热浪天气下局地尺度热岛特征研究

Characterization of the Urban Heat Island at the Local Scale during Heatwaves in Hot and Humid City

陈漪淇，陈光，何昊，谭啸，蔡云楠
Chen Yiqi, Chen Guang, He Hao, Tan Xiao, Cai Yunnan

摘要 城市化进程中，城市热岛（UHI）效应和高温热浪（HW）事件频率增加，对人类健康和城市生态系统影响巨大。基于局地气候分区（LCZ）框架，研究 2021 年广州的 HW 事件对局地 UHI 的影响。结果表明，LCZ_1,2,3,4 在 HW 期间城市热岛强度（UHII）比非热浪（NHW）期间明显增强，且夜间增强大于白天，平均放大了 0.1~0.6℃，显示出 UHI 与 HW 之间的协同效应；LCZ_A 在夜间平均 UHII 增加 0.12℃，但白天冷岛效应增强。HW 期间强太阳辐射和低风速使 LCZs 的 UHII 提升，日最大热岛强度平均值提升 0.84℃。

关键词 城市热岛；高温热浪；局地气候分区；城市形态；日最大热岛强度

Abstract Urbanisation has increased the frequency of urban heat island (UHI) effects and heat wave (HW) events, with significant impacts on human health and urban ecosystems. Based on the framework of Local Climate Zones (LCZs), this study investigated the impact of HW events on the UHI at the local scale of Guangzhou in 2021. The results showed that LCZ_1,2,3,4 have a significantly stronger urban heat island intensity (UHII) during HW than during non-heat wave (NHW), and the enhancement is greater at night than during the day, with an average amplification of 0.1~0.6°C, showing a synergistic effect between UHI and HW. LCZ_A has an average UHII increase of 0.12℃ at night, but the urban cold island effect is enhanced during the day. Strong solar radiation and low wind speeds increased UHII in LCZs during HW, and the average daily UHII_max increased by 0.84°C.

Keywords Urban heat island; Heat wave; Local climate zone; Morphological parameters; Daily maximum heat island intensity

陈漪淇，广东工业大学，硕士研究生。
陈光，广东工业大学，讲师。
何昊，广东工业大学，硕士研究生。
谭啸，广东工业大学，硕士研究生。
蔡云楠，广东工业大学，教授。

城市立交桥下空间热环境研究——以成都市府青路立交桥下空间为例

Research on Thermal Environment of Space Under Urban Overpass —Taking Space Under Fuqing Road Overpass in Chengdu as an Example

陈姝亚，高庆龙，欧阳金龙
Chen Shuya,
Gao Qinglong,
Ouyang Jinlong

摘要 城市立交桥下空间常被作为休闲活动空间。由于所处区位、使用功能及使用方式的不同，其空间环境具有较大差异。研究选取成都市在利用的立交桥下空间，采用热环境现场实测及行为注记法，基于空间不同热环境特征，分析空间人体热舒适度、使用者行为特征及环境偏好。结果表明，热舒适受太阳辐射影响较大，桥面的遮蔽效果能够显著提升人体热舒适度，而依据随太阳运动轨迹变化的热环境，匹配强度适宜的活动，能优化立交桥下空间功能配置与分布，助力桥下空间使用率提升、环境品质改善。

关键词 立交桥下空间；热环境；行为偏好；提升策略

Abstract The spaces under urban overpass are often used as spaces for leisure activities. Due to the different locations, functions and so on, the spatial environments are quite different. The study adopts thermal environment field measurements and behavioral notation method to analyze human thermal comfort, behavioral characteristics and environmental preferences of the utilized spaces under overpass in Chengdu, based on thermal environment characteristics. The results show that thermal comfort is greatly affected by solar radiation, and the shading effect of overpass significantly improves human thermal comfort. Based on the thermal environment that changes with the sun's trajectory, matching activities with appropriate intensity can optimize the functional configuration and distribution. It helps to increase the utilization rate and improve the environmental quality of space under overpass.

Keywords Space under overpass; Thermal environment; Behavioral preferences; Enhancement strategies

陈姝亚，四川大学，硕士研究生。
高庆龙，四川大学，博士，教授。
欧阳金龙，四川大学，博士，教授。

基于室外热舒适和能耗的住区布局形态综合评价研究

Comprehensive Evaluation of Residential Layout Form Based on Outdoor Thermal Comfort and Energy Consumption

徐燊，王思尧
Xu Shen,Wang Siyao

摘要 居住街区是人们活动的主要场所，其经济性和舒适性至关重要。现有形态学研究多关注于单一性能，而忽略了室外热舒适和能耗双重视角下的形态优化研究。因此，本文以武汉市住区为研究对象，通过建立综合性能评价标准，探究 7 种住区布局形态对室外热舒适和能耗的影响差异。研究结论：①各布局形态的制冷能耗与采暖能耗、冬季、夏季室外热舒适存在差异；②点式交错式和点式周边式的综合表现更佳。

关键词 住区；形态；室外热舒适；能耗；数值模拟；综合评价

Abstract Residential neighborhood is the main place for people's activities, and its economy and comfort are crucial. Most existing morphological studies focus on a single performance, but ignore the study of morphological optimization under the dual perspective of outdoor thermal comfort and energy consumption. Therefore, this paper investigates the differences of seven settlement layout morphologies on outdoor thermal comfort and energy consumption by establishing comprehensive performance evaluation criteria, taking Wuhan city settlements as the research object. The study concludes that: ① there are opposite differences between cooling energy consumption and heating energy consumption, outdoor thermal comfort in winter and summer for each layout form; ② the point staggered type and point perimeter type have better integrated performance.

Keywords Residential neighborhoods; Morphology; Outdoor thermal comfort; Energy consumption; Numerical simulation; Comprehensive evaluation

徐燊，华中科技大学建筑与城市规划学院，教授。
王思尧，华中科技大学建筑与城市规划学院，硕士研究生。

重庆某大学校园夏季避暑地图研究

Cooling Walking Map of a University Campus in Chongqing in Summer

张海滨，尹婧，周千叶，徐岳荣，汤加
Zhang Haibin, Yin Jing, Zhou Qianye, Xu Yuerong, Tang Jia

摘要 近年来，城市高温热浪频现，校园的热环境对师生的健康出行具有重要影响。本研究以重庆市某校园为研究对象，采用问卷调研和现场测试的方法，分析温度、湿度等气候要素的空间分布情况和时间变化规律，归纳出影响校园热环境的空间要素及校园夏季热环境气候分布特征，以热舒适为目标，绘制夏季校园避暑地图，为师生的夏季出行活动提供参考。研究表明，在校园环境设计过程中，合理控制植被、道路、铺地、水景的位置和面积，可以为人们在校园内的活动提供避暑降温的便利，营造舒适的校园室外场所。

关键词 重庆；大学校园；避暑地图；热舒适；热环境

Abstract In recent years, heat waves have occurred frequently in cities, and the thermal environment of the campus has an important impact on the health of teachers and students. This study takes a campus in Chongqing as the research object, adopting questionnaire survey and field test, to analyze the spatial distribution and temporal variation of temperature and humidity, and to summarize the spatial elements that affect the thermal environment of the campus and the distribution characteristics of the campus' s thermal environment in summer. With the goal of thermal comfort, a campus cooling walking map is established to provide reference for teachers and students' summer outdoor activities. The study shows that reasonable design of the location and area of vegetation, roads, paving and water features can provide people with the cooler environment for activities on campus and create a comfortable outdoor campus in the process of campus environment design.

Keywords Chongqing; University campus; Cooling walking map; Thermal comfort; Thermal environment

张海滨，重庆大学建筑城规学院，副教授，建筑技术科学系副主任。
尹婧，周千叶，徐岳荣，汤加，重庆大学建筑城规学院，本科生。

城市环境下的室外温度修正方法研究——以西安为例

A Study on Outdoor Temperature Correction Method in Urban Environment —A Case Study of Xi 'an

吕文，李红莲，宋学锟，和玺，霍英博
Lü Wen, Li Honglian, Song Xuekun, He Xi, Huo Yingbo

摘要 随着我国“双碳”目标与城市化进程的不断推进，使得城市不同位置的气象数据需求也在不断增加。目前气象观测站一般离城市中心距离较远，气象要素观测数量有限，无法做成建筑能耗分析直接可用的天气文件，不能满足建筑节能设计所需。本文基于西安市气象观测站与区域站数据，结合气象参数水平维度上的变化规律，计算出温度取值区间，结合多模型对温度进行修正，并利用 EnergyPlus 进行能耗模拟分析。结果表明使用经过多模型修正后的 TMY 得到的模拟结果非常接近该地的实际能耗。

关键词 气候模型；温度分布；建筑能耗；典型气象年

Abstract With the continuous advancement of China's carbon peaking and carbon neutrality goals and urbanization process, the demand for meteorological data in different locations within cities is also increasing. Currently, meteorological observation stations are generally located far from city centers, and the limited number of observed meteorological variables makes it impossible to create weather files directly usable for building energy analysis, thus failing to meet the requirements for energy-efficient building design. This paper is based on the meteorological data from both the meteorological observation station and regional station in Xi'an. By considering the variation patterns of meteorological parameters along the horizontal dimension, the temperature range is calculated. Furthermore, temperature is adjusted using multiple models, and energy simulation analysis is conducted using EnergyPlus. The results show that the simulation results obtained using the TMY after being corrected by multiple models are closer to the actual energy consumption of the location.

Keywords Climate models; Temperature distribution; Building energy consumption; TMY

吕文，西安建筑科技大学，硕士研究生。
李红莲，西安建筑科技大学，教授。
宋学锟，和玺，霍英博，西安建筑科技大学，硕士研究生。

厦门冬春季居住区户外体育空间热舒适研究

Study on Xiamen' s Winter and Spring Thermal Comfort of Outdoor Sports Space in the Residential Community

叶静怡，吴正旺，鲍沁雨
Ye Jingyi,
Wu Zhengwang,
Bao Qinyu

摘要 本研究以厦门为例，选取户外体育空间场地较为集中、设施功能较丰富的居住区为研究对象，通过场地观测、微气候数据采集、问卷调查和居民访谈等方法，深入调查冬春季居住区户外体育空间的小气候特征，探究各种类型户外体育活动的人体舒适度与小气候要素之间的复杂关系，以提出改善湿热地区冬春季户外体育空间热舒适的设计方法。结果表明，气温、太阳辐射强度和空气湿度是影响居民开展户外体育活动热舒适的最主要因素，随着体力活动强度上升，认为气温和太阳辐射为影响舒适度主要因素的人群比例下降。

关键词 户外体育空间；厦门冬春季；居住区；户外热舒适；湿热地区

Abstract Taking Xiamen as an example, this study selects a residential community with clustered outdoor sports venues and rich facilities as the research object. Through methods such as site observation, microclimate data collection, questionnaire survey and resident interviews, the microclimate characteristics of outdoor sports spaces in the residential community in winter and spring are thoroughly investigated, and the complex relationship between human comfort and microclimate factors in various types of outdoor sports activities is explored, in order to propose design methods for thermal comfort of outdoor sports spaces in spring, in order to improve winter and spring comfort in humid subtropical areas. The results show that air temperature, solar radiation intensity and air humidity are the main factors affecting the thermal comfort of outdoor sports activities. As the intensity of physical activity increases, the proportion of people who believe that temperature and solar radiation are the main factors affecting comfort decreases.

Keywords Outdoor sports space; Xiamen winter and spring; Residential community; Outdoor thermal comfort; Humid subtropical areas

叶静怡，华侨大学，建筑学博士研究生。
吴正旺，华侨大学，建筑学教授，博士生导师。
鲍沁雨，南京大学城市规划设计研究院有限公司，助理规划师。

垂直绿化街区行人层面降温增湿的影响研究

Study on the Influence of Cooling and Humidification at Pedestrian Level in a Vertical Greening Block

张海滨，柳佩渝，陈舒，孙佳奇
Zhang Haibin, Liu Peiyu,
Chen Shu, Sun Jiaqi

摘要 为研究垂直绿化对街区层面降温增湿的影响，在夏季对重庆外滩广场的垂直绿化街区进行热环境实测，分析比较了叶面积指数、绿化高度、距外墙距离等因素对绿墙降温增湿能力的影响。实验结果表明 12m 高的绿墙相比无绿化墙体可降低平均温度 1.37℃；叶面积指数较大的 12m 高绿墙可实现平均温度降低 3.05℃。绿墙的高度及植物的叶面积指数与其降温增湿能力正相关。本文研究结果可促进实现改善城市热环境的目标，为改善街区尺度热环境的垂直绿化设计提供指导。

关键词 垂直绿化；降温增湿；叶面积指数；绿墙高度

Abstract In order to study the influence of vertical greening on the cooling and humidifying of the heart heat level of the block, the thermal environment of a 450 m-long vertical greening block in the Bund Square in Chongqing was measured in hot summer, and the effects of leaf area index, greening height and distance from the outer wall on the cooling and humidifying capacity of the green wall were analyzed and compared. The experimental results show that the average temperature of the 12m-high ordinary green wall can be reduced by 1.37℃ .The 12m high green wall with large leaf area index can reduce the average temperature by 3.05℃ . The height of the plant wall and the leaf area index of the plant are correlated with the cooling and humidifying capacity of the green wall. The research results are expected to promote the large-scale construction of vertical greening, achieve the goal of improving the urban thermal environment and provide guidance for the vertical greening design of improving the thermal environment at the block scale.

Keywords Vertical greening; Thermal environment; Leaf area index; Green wall height

张海滨，重庆大学建筑城规学院，副教授。
柳佩渝，重庆大学建筑城规学院，硕士研究生。
陈舒，重庆大学建筑城规学院，硕士研究生。
孙佳奇，重庆大学建筑城规学院，硕士研究生。

一种低空双源遥感城市热环境观测平台的性能分析

Performance Analysis of a Low-Altitude Dual-Source Remote Sensing Platform for Urban Thermal Environment Observation

袁旭，韩家亮
Yuan Xu，Han Jialiang

摘要 随着全球变暖与城市化进程的加速，城市的热环境问题日益突出。传统的卫星城市热环境遥感存在着空间分辨率低、回访周期长等缺陷。低空热红外遥感尽管能够弥补以上缺陷，然而数据单一，无法对下垫面精准分类，难以展开深入分析。针对上述问题，我们拟采用无人机搭载热红外和高光谱传感器，设计一个质轻、便捷、集成度高、稳定性好的低空双源遥感城市热环境观测平台。本文对该平台的组成与功能进行了介绍，分析了平台的性能，对比其他的城市热环境观测平台，对其优势进行了阐述。本文对城市热环境的监测具有推动作用。

关键词 城市热环境；高光谱；热红外；无人机；低空

Abstract With the acceleration of global warming and urbanisation, the problem of urban thermal environment has become increasingly prominent. Traditional satellite remote sensing of urban thermal environment has shortcomings such as low spatial resolution and long return visit period. Although low-altitude thermal infrared remote sensing can compensate for the above shortcomings, the data is not sufficient to accurately classify of the underlying surface, making it difficult to conduct in-depth analysis. Aiming at the above problems, we plan to develop a low-altitude dual-source remote sensing platform for urban thermal environment observation, using an unmanned aerial vehicle (UAV) equipped with a thermal infrared sensor and a hyperspectral sensor. This platform is light, convenient, highly integrated, and stable. In this paper, we introduce the composition and functions of the platform, analyse the performance of the platform, and explain its advantages compared with other urban thermal environment observation platforms. This study can promote the monitoring of urban thermal environment.

Keywords Urban thermal environment; Hyperspectral; Thermal infrared; UAV; Low altitude

袁旭，郑州大学建筑学院 / 华南理工大学亚热带建筑与城市科学全国重点实验室，讲师。
韩家亮，郑州大学建筑学院，硕士研究生。

寒地校园步行道路秋季视觉环境因子与热舒适相关性研究

Correlation Analysis Between Fall Visual Environment Elements and Thermal Comfort of Campus Walkway In Cold Regions

黄锰，刘丽明
Huang Meng, Liu Liming

摘要 本文为三维空间中的步行环境与人体热舒适的相关性研究，以哈尔滨工业大学六条步行道路为研究对象，将道路视觉环境的组成要素量化为视觉环境因子，并进行数据提取；采用实测 + 问卷 + 计算分析的主客观研究方法，探究视觉环境因子与热舒适、整体舒适度相关性。研究显示：热舒适、舒适度与视觉环境因子相关性一致，其中与绿化、天空、道路因子呈正相关，与建筑、车辆呈负相关；不同视觉环境因子与热舒适、舒适度相关强度不同，绿化、建筑因子是关键因素；热舒适与视觉舒适度、整体舒适度呈正相关。

关键词 步行道路；热舒适；视觉环境；深度学习

Abstract This paper investigates the relationship between the walking environment and human thermal comfort in three dimensions, and takes six walkways at Harbin Institute of Technology as the research object, quantifies the road visual environment components into visual environment elements, and extracts the data. The study investigates the relationship between visual environment elements and thermal comfort and comfort by using subjective and objective research methods of real measurement, questionnaire, and mathematical analysis. The research shows that thermal comfort and overall comfort are consistent with visual environment elements, which are positively correlated with green, sky and road and pavement elements, and negatively correlated with construction and vehicle elements. Different visual environment elements have varying degrees of association with thermal comfort and overall comfort, with green and construction being the most important. Thermal comfort and holistic comfort are positively related to with visual comfort.

Keywords Walkway; Thermal comfort; Visual environment; Deep learning

黄锰，哈尔滨工业大学建筑学院 / 寒地城乡人居环境科学与技术工业和信息化部重点实验室，副教授，硕士生导师。
刘丽明，哈尔滨工业大学建筑学院，硕士研究生。

基于广场空间热应力水平的树木降温效益评估

Evaluation of the Cooling Benefits of Trees Based on the Thermal Stress in the Square

郭若男，郭飞，郑如薇，董菁，赵君

Guo Ruonan, Guo Fei, Zheng Ruwei, Dong Jing, Zhao Jun

郭若男，大连理工大学建筑与艺术学院，博士研究生。
郭飞，大连理工大学建筑与艺术学院，教授，博士生导师，副院长。
郑如薇，大连理工大学建筑与艺术学院，硕士研究生。
董菁，大连理工大学建筑与艺术学院，博士后。
赵君，大连理工大学建筑与艺术学院，博士研究生。

摘要 在气候变暖背景下进行适应性城市设计是当下的重要议题。相较于城市行道树的精准规划，很少有针对城市广场提供高效遮阴的实用性研究。本文基于广场空间提出了降低热风险的方法，引入反映人类热应激的热应力强度指标。基于占用最少空间的前提，在典型广场开展树冠降温效益评估，以确定树木降温的影响因素、不同广场的降温效果以及缓解路径。研究表明，冠层大小是决定因素，落叶树和针叶树在种植距离和降温效果上存在差异。小尺度广场易实现整个区域的改善，大尺度广场在其空间边界进行种植是大幅缓解局部热应力的经济性策略。

关键词 气候变暖；热风险；降温；树种；广场

Abstract Adaptive urban design is an important issue in the context of climate warming. Compared to the meticulous planning of street trees in urban areas, there is a lack of practical studies on the effective provision of shade in urban squares. This paper presents a proposed method to reduce the level of thermal risk with a focus on squares. The method introduces the heat stress intensity index, which reflects the level of heat stress experienced by humans. Based on the premise of occupying the least space, an assessment of tree canopy cooling benefits is conducted in typical squares. The objective is to determine the factors that influence tree cooling, the cooling effect of different squares, and mitigation strategies. The study shows that canopy size is the determining factor, and there are differences between deciduous and coniferous trees in terms of planting distance and their cooling effect. Small-scale squares are easy to achieve the improvement of the whole area, while large-scale squares with vegetation along their boundaries are cost-effective strategies to significantly reduce local heat stress.

Keywords Climate warming; Thermal risk; Cooling; Tree species; Squares

---

校园绿化对区域热环境影响的 UWG 模拟与分析

UWG Simulation of the Impact of Campus Greening on the Regional Thermal Environment

何西子，傅秀章

He Xizi，Fu Xiuzhang

何西子，东南大学建筑学院，硕士研究生。
傅秀章，东南大学建筑学院，副教授。

摘要 受城市热岛效应影响，城镇之间存在温度差，通过提高绿化覆盖率与降低建筑密度的措施会降低热岛强度。本文选取城市范围内的校园空间作为研究对象，基于 Grasshopper 下的城市气象生成器插件模拟分析东南大学三个校区的微气候环境；比较各校区不同绿化覆盖率和建筑密度对微气候的影响。结果显示，三个校区的城市气象数据与南京气象站气候数据呈高相关性，但仍存在 0.82~−1.96℃温差。三个校区因绿化率的不同温湿度也存在差异，其中绿化率最高的九龙湖校区最接近气象站数据。

关键词 城市热岛；城市气候；UWG；天气数据；校园空间

Abstract Due to the urban heat island phenomenon, cities exhibit temperature differences that can be mitigated by increasing green cover. This study focuses on the campus spaces in urban areas, specifically three sites within Southeast University. Using the Urban Weather Generator (UWG) plug-in under Grasshopper simulates and analyzes the microclimate of each campus to examine the impact of incorporating greenery with varying greening rates. The result indicates that the urban weather data of the three campuses are highly correlated with the climate data of the Nanjing weather station, differing by only about 0.82~−1.96℃. Furthermore, the temperature and humidity at each location were affected differently by the greening rate, with the Jiulong Lake campus exhibiting the greatest resemblance to data from the weather station, due to its elevated green coverage.

Keywords Urban heat island; Urban climate; UWG; Weather data; Campus space

“双碳”战略视角下寒地住区公共空间微气候适应性提升策略

Strategy for Improving the Microclimate Adap-tability of Residential Public Spaces in Cold Regions from the Perspective of “Carbon Peaking and Carbon Neutralization Strategy”

陈晓丹，林春翔 *，梅兰
Chen Xiaodan, Lin Chunxiang, Mei Lan

陈晓丹，河北建筑工程学院建筑与艺术学院，讲师。
林春翔，北京交通大学土木建筑工程学院，博士研究生。
梅兰，河北建筑工程学院建筑与艺术学院，讲师。

摘要 面对全球气候变化和能源可持续发展的需求，科学系统地提高寒冷地区住区公共空间微气候适应性，将为碳源减排与碳汇增容提供有力支持。首先，本文基于寒地气候特征与住区公共空间现状，提出减碳增汇途径。同时，以具体实例和数据支撑，多角度分析寒地住区微气候环境与“双碳”目标的关系。最后，提出“双碳”战略视角下的微气候适应性提升策略，即经济效益需求转向绿色可持续开发、平面扩张转向地上地下协调开发、单一资源开发转向统筹复合利用。

关键词 “双碳”战略；住区公共空间；寒冷地区；微气候；适应性；提升策略

Abstract In response to the needs of global climate change and sustainable energy development, scientific and systematic improvement of microclimate adaptation in public residential spaces in cold regions will provide strong support for carbon emission reduction and carbon sink capacity expansion. First, based on the climate characteristics of cold regions and the current spatial situation of residential spaces, two ways of reducing carbon and increasing sinks are proposed. At the same time, supported by specific examples and data, the relationship between the microclimate environment of the cold region residential space and the strategic goals of “carbon peaking and carbon neutralization” is analyzed from differrent perspectives. Last, the enhancement strategies for microclimate adaptation in the view of “carbon peaking and carbon neutralization” are put forward, including shifting demand for economic efficiency to green and sustainable development, switching from surface expansion to development to comprehensive utilization.

Keywords Carbon peaking and carbon neutralization strategy; Residential public space; Cold regions; Microclimate; Adaptability; Enhancement strategies

---

基于风环境的寒冷地区高校生活区绿化优化设计

Optimization Design of Greening in Universities Residential Area in Cold Regions Based on Wind Environment

叶萌，陈卓伦，郭嘉
Ye Meng, Chen Zhuolun, Guo Jia

叶萌，华南理工大学建筑学院，硕士研究生。
陈卓伦，华南理工大学建筑学院，高级工程师。
郭嘉，华南理工大学建筑学院，高级工程师，工作二室主任。

摘要 合理的风环境设计对高校师生的校园生活质量和身心健康有着积极作用。本文以中国民航大学二期生活区规划设计方案为例，运用 CFD 模拟的方法分析生活区室外风环境，探讨不同绿化布局和植物组合对典型区域人行层风环境的影响，提出针对性优化策略，后对生活区优化方案再次模拟，模拟结果表明优化后场地内风环境舒适度有明显提高，合理的绿化设计能够改善生活区风环境。

关键词 寒冷地区；高校生活区；风环境；绿化优化设计

Abstract The reasonable design of wind environment in universities plays a positive role in the quality of campus life and the physical and mental health of teachers and students. Taking the second phase planning design of the residential area of the Civil Aviation University of China as an example, this paper uses CFD simulation method to analyze the outdoor wind environment of the residential area. It investigates the impact of green layout on the pedestrian wind environment in typical windward area, and proposes targeted optimization strategies. The simulation results show that the comfort of the wind environment has been significantly improved, and reasonable greening design can improve the wind environment in the residential area.

Keywords Cold regions; Universities residential area; Wind environment; Greening optimization design

圆形土楼在重建时的夏季室内风环境模拟研究

Simulation Study on the Summer Indoor Wind Environment of Circular Tulous Building during Reconstruction

赖婉玲，林姝颖，吴正旺
Lai Wanling, Lin Shuying, Wu Zhengwang

摘要 在乡村振兴和“双碳”政策背景之下，使用现代建筑技术及材料对损毁的非历史保护性土楼进行生态重建，对于土楼的传承有重要意义。为探究重建过程中由于材料、结构、空间不同导致的室内风环境变化，本文选取圆形土楼作为研究对象，基于福建漳州地区的地域气候特征，采用 Phoenics 2019 和 Airpak 3.0 版软件进行数值模拟发现：①东西侧用房的室内通风效果最差，西侧尤甚；②单元式圆形土楼室内通风效果最好；③坡屋顶有利于室内的自然通风，设置拔风烟囱能增强其通风效果。

关键词 福建土楼；室内风环境；数值模拟；重建

Abstract Under the policy background of rural revitalization and “carbon peaking and carbon neutrality”, it is of great significance for the inheritance of ecological and cultural value of Tulous to use modern architectural technology and materials to carry out ecological reconstruction of damaged non-historic protective Tulous. In order to explore the indoor ventilation changes caused by different materials, structures and spaces during the reconstruction process, this paper selects the circular Tulous as the basic prototype, based on the regional climate characteristics of Zhangzhou, Fujian, and uses Phoenics 2019 and Airpak 3.0 software. Through numerical simulation, it is found that: ① The indoor ventilation effect of the room on east-west side is the worst, especially on the west side; ② The indoor ventilation effect of the unit circular earth building is the best; ③ The sloping roof is conducive to natural indoor venti-lation, and the ventilation effect can be improved by building up the chimney.

Keywords Fujian Tulous; Indoor wind environment; Numerical simulation; Reconstruction

赖婉玲，华侨大学建筑学院，硕士研究生。
林姝颖，华侨大学建筑学院，硕士研究生。
吴正旺，华侨大学建筑学院，教授。

南京江北新区局地气温时空变化特征研究

Analysis of Temporal and Spatial Characteristics of Local Temperatures in the Jiangbei New Area of Nanjing

徐沙沙，杨小山
Xu Shasha, Yang Xiaoshan

摘要 局地气候分区体系可用于分析城市局地气温与城市空间形态间的关系，揭示城市内部热岛分布特征。本研究基于 LCZ 体系在南京江北新区选取了 50 个典型 LCZ 地块，采用固定测点方式进行长期气温观测。从时间变化和空间分布两个角度分析了 2021 年 5 月至 2022 年 5 月不同类型 LCZ 地块的气温时空变化特征。结果表明，时间变化上，不同 LCZ 地块的热岛强度呈现显著差异，总体表现为“昼弱夜强、夏弱冬强”的特征；相同 LCZ 类型地块受地理位置和下垫面材质构造的影响，存在热岛强度差异。

关键词 城市热岛；城市冠层；局地气候分区；气温观测

Abstract Local Climate Zones (LCZ) can be used to analyze the relationship between urban air temperature and urban morphology, revealing the spatial variability of urban heat islands. Based on the LCZ system, 50 plots were selected in Nanjing Jiangbei New Area. Fixed stations were set up to monitor the long-term temperature. We analyzed the temporal and spatial variation characteristics of different types of LCZs for the period from May 2021 to May 2022. The results showed that the urban heat island intensity (UHII) of different LCZs showed significant differences in time variation. In general, there was a pattern of weak daytime UHII and strong nighttime UHII and weak summer UHII and strong winter UHII was displayed. In terms of spatial distribution, there were also differences in UHII between the sites belonging to the same LCZ type, which can be attributed to the influence of geographical location and sitecharacteristics.

Keywords Urban heat island; Urban canopy layer; Local climate zone; Air temperature measurement

徐沙沙，南京工业大学，硕士研究生。
杨小山，南京信息工程大学，教授。

广州大气 NO_2 浓度与气象参数关系的量化研究

Quantitation on the Relationship Between Atmospheric NO_2 Concentration and Meteorological Parameters in Guangzhou

颜艺璇，任鹏
Yan Yixuan, Ren Peng

摘要 当排放源相对稳定时，气象参数会在一定程度上决定大气污染物的时空分布特征。本文分析了 2015—2019 年广州市 NO_2 浓度的变化特征，以同期逐日气象参数作为解释变量建立了 NO_2 浓度的广义加性模型（GAM），解析并量化了影响 NO_2 浓度的主要参数及其响应关系。研究发现，2015—2019 年广州大气 NO_2 浓度总体呈上升趋势，并表现出夏低冬高的季节性特征。NO_2 浓度的 GAM 的决定系数 R_2 和方差解释率分别为 0.890 和 89.3%，其中气象变量的影响占 41.2%。另外，响应图揭示了 NO_2 浓度与气象变量之间复杂的非线性关系，并为后续研究影响其他污染物的气象参数提供了坚实的基础。

关键词 NO_2 浓度；气象参数；广义加性模型；广州

Abstract When the emission source is relatively stable, meteorological factors will influence the temporal and spatial distribution of air pollutants to a certain extent. This paper first analyzed the change characteristics of NO_2 concentration in Guangzhou from 2015 to 2019, and then established a generalized additive model (GAM) of NO_2 with daily meteorological parameters as explanatory variables to quantify the influence of these main parameters on NO_2 concentration and their response relationships. The determination coefficient R_2 and variance explanation rate of GAM were 0.890 and 89.3%, respectively, and the total impact of the meteorological variables accounted for 41.2%. In addition, the response plots revealed complex nonlinear relationships between NO_2 concentrations and the meteorological variables, and provided a solid basis for subsequent studies of meteorological parameters affecting other pollutants.

Keywords NO_2 concentration; Meteorological parameters; Generalized additive model; Guangzhou

颜艺璇，华南理工大学建筑学院，硕士研究生。
任鹏，华南理工大学建筑学院，亚热带建筑与城市科学全国重点实验室，副教授，硕士生导师。

基于实证方法的城市绿化因子与微气候的相关机制研究

Research on the Correlation Mechanism Between Urban Greening Factors and Microclimate Based on Empirical Methods

魏瑞涵，徐畅，宋德萱 *
Wei Ruihan, Xu Chang, Song Dexuan

摘要 城市绿化与城市微气候之间有着密切的关系。通过实地获取监测上海地区大学校园内春夏两季晴天的环境气象参数；通过实地调研、无人机拍摄等多元途径获取并计算绿容率（GnPR）、乔木树冠投影面积（Tree）、灌木投影面积（Shrub）、草坪面积（Grass）等绿化参数；利用鱼眼镜头法获取天空视域因子（SVF）。利用以上几组数据分析了不同绿化因子对微气候影响的差异，结果表明绿化能有效调节气候，落叶乔木降低空气温度和太阳辐射的能力最强。研究结果为环境微气候的城市设计提供了理论基础和实践指南。

关键词 城市绿化因子；城市微气候；热舒适度评价；相关性机制

Abstract There is a close relationship between urban greening and urban microclimate. This study obtained monitoring data from the field of environmental meteorological parameters in the spring and summer seasons of sunny days on the campus of universities in Shanghai; Green parameters such as green volume ratio (GnPR), tree canopy projection area (Tree), shrub projection area (Shrub), and lawn area (Grass) were obtained and calculated through multiple methods such as field research and drone shooting. The Sky Field of View Factor (SVF) was obtained by using the fisheye lens method. Using the above sets of data analyzed the differences in the impact of different greening factors on microclimate. The results indicated that greening can effectively regulate the climate, and deciduous trees have the strongest ability to reduce air temperature and solar radiation. The research results could provide a theoretical basis and practical guide for urban design of environmental microclimate.

Keywords Urban greening factor; Urban microclimate; Thermal comfort evaluation; Correlation mechanism

魏瑞涵，山东建筑大学建筑城规学院，讲师。
徐畅，山东建筑大学建筑城规学院，硕士研究生。
宋德萱，同济大学建筑与城市规划学院，教授，建筑环境梯队负责人。

德格印经院风热环境实测与数字化仿真

Measurement and Digital Simulation of the Wind-thermal Environment of Dege Scripture Printing House

王晓亮，钟磊，李姝靓，麦贤敏，魏庆芃
Wang Xiaoliang, Zhong Lei, Li Shuliang, Mai Xianmin, Wei Qingpeng

王晓亮，西南民族大学建筑学院，副教授。
钟磊，西南民族大学建筑学院，硕士研究生。
李姝靓，西南民族大学建筑学院，讲师。
麦贤敏，西南民族大学建筑学院，教授，院长。
魏庆芃，清华大学建筑学院，副教授。

摘要 德格印经院作为藏式木构夯土建筑的典型代表，在应对川西高原冬季寒冷、昼夜温差大和太阳辐射强等特殊气候的过程中，形成了独具特色的建筑气候适应性营造技艺，对文物保护和文化传承起到了重要作用。为科学认识德格印经院的通风现状及风热环境规律，采用现场实测和 CFD 数字化仿真相结合的方法对德格印经院的热湿环境、通风特征及气流组织进行研究。结果表明，德格印经院已形成合理的热湿环境，有良好的室内通风效果，对川西高原传统建筑通风及热湿环境的营造具有重要的借鉴意义。

关键词 德格印经院；现场实测；CFD 仿真；风环境；热环境

Abstract As a typical representative of Xizang-style wooden rammed earth architecture, Dege Scripture Printing House has developed a unique building climate adaptation technique in the process of dealing with the special climate of the western Sichuan Plateau, such as cold winter, large temperature difference between day and night and strong solar radiation, which plays an important role in the protection of cultural relics and cultural inheritance. In order to understand the ventilation status and law of Dege Scripture Printing House scientifically, the thermal environment, ventilation characteristics and air distribution of Dege Scripture Printing House were studied by combining field measurement and CFD digital simulation. Results show that Dege Scripture Printing House has formed reasonable thermal environment and good indoor ventilation effect, which has important reference significance for the construction of the ventilation and thermal environment of traditional buildings in western Sichuan Plateau.

Keywords Dege Scripture Printing House; Field measurement; CFD simulation; Wind environment; Thermal environment

---

基于天空开阔度的城市局地气候区研究

Study on the Typologies of Local Climate Zones Based on the Sky View Factor

杜思宏，石邢
Du Sihong, Shi Xing

杜思宏，同济大学建筑与城市规划学院，博士研究生。
石邢，同济大学建筑与城市规划学院，教授，博士生导师。

摘要 构建城市形态与热环境的关联性是优化城市热环境的基础，经典局地气候区的 10 种街区类型为研究城市热环境特征提供了有效方法，但在天空开阔度范围方面还无法涵盖城市的典型街区类型。本文以上海为例，根据外环内街区的建筑密度、平均建筑高度与天空开阔度建立了 32 个典型街区，模拟了各街区的热环境，并使用聚类分析方法将街区分类。结果表明，建筑密度及高度均不变，天空开阔度的差异依然会使街区形成明显不同的热环境，根据街区形态与热环境综合特征可分类获得 14 种街区原型，拓展了经典局地气候区体系。

关键词 城市热环境；局地气候区；街区原型；聚类分析；天空开阔度

Abstract Developing the correlation between urban morphology and the thermal environment is essential for improving the urban thermal environment. The classic Local Climate Zone (LCZ) scheme provides an efficient method to investigate urban thermal performances. However, the ranges of sky view factor (SVF) of 10 LCZ built types are still unable to cover those of typical blocks in contemporary cities. This study simulated the thermal environments of 32 prototypical blocks, which were built according to the building footprint ratios (BFRs), mean building heights (MBHs), and mean SVFs of actual blocks in Shanghai, China. Then clustering analysis was used to classify the 32 blocks. Results show that when the BFR and MBH are constant, SVF still contributes to distinctly different thermal environments for urban blocks. The prototypical blocks can be classified into 14 types, which extends the classic LCZ scheme.

Keywords Urban thermal environment; Local Climate Zone; Urban blocks; Clustering analysis; Sky view factor

树木垂直结构对竖向温度的影响研究

Influence of Tree Vertical Structure on Vertical Temperature

王薇，何宝杰
Wang Wei, He Baojie

摘要 树木对于调节小气候、缓解热岛和改善人类热舒适非常重要。研究表明，树木引起的降温效益与树冠形态、下垫面和周围环境有关。然而，目前大多数研究都集中探讨 1.5 m 高度处的微气候，而缺乏对于树木垂直温度变化的理解。因此，本文旨在分析行道树常见的两种树木配置形式的垂直冷却性能。该实验在重庆市进行，现场实验选取了树 + 草（组合 A）和树 + 灌木（组合 B）作为实验对象，分别测量 0.5 m、1.0 m、1.5 m、2.0 m 和 2.5 m 处的垂直空气温度。结果显示，组合 A 的最大垂直温差为 1.01℃，发生在 1.5 m 和 2.0 m 之间。组合 B 的最大垂直温差为 1.0 ℃，发生在 2.5 m 和 2.0 m 之间。

关键词 树木配置；竖向温度分布；树木降温；微气候

Abstract Trees are important for regulating microclimates, mitigating heat islands and improving human thermal comfort. Research has shown that tree-induced cooling benefits are related to canopy morphology, surface cover and surroundings. However, most current studies have focused on microclimates at 1.5 m height and lack an understanding of vertical temperature variation in trees. Therefore, this study analyzed the vertical cooling performance of two forms of tree configuration commonly found in street trees. The experiment was conducted in Chongqing, China, and selected to measure vertical air temperature and wind speed at 0.5 m, 1.0 m, 1.5 m, 2.0 m and 2.5 m for tree + grass (ComA) and tree + shrub (ComB). The results showed that the maximum vertical temperature difference for Com A was 1.01°C and occurred between 1.5 m and 2.0 m. The maximum vertical temperature difference for ComB was 1.0°C and occurred between 2.5 m and 2.0 m.

Keywords Tree configuration; Vertical temperature distribution; Tree cooling; Microclimate

王薇，重庆大学建筑城规学院，气候韧性与低碳城市研究中心，重庆大学溧阳智慧城市研究院，硕士研究生。
何宝杰，重庆大学建筑城规学院，气候韧性与低碳城市研究中心，重庆大学溧阳智慧城市研究院，研究员。

基于太阳辐射的夏热冬冷地区街区形态优化设计研究

Optimization Design of Block Form in Hot Summer and Cold Winter Areas Based on Solar Radiation

徐燊，侯俊豪，谢孟举，桑梦城
Xu Shen, Hou Junhao, Xie Mengju, Sang Mengcheng

摘要 夏热冬冷地区由于气候的特殊性，夏季制冷能耗高，冬季采暖能耗高。夏热冬冷地区建筑太阳能辐射的利用对于改善室内热环境、降低建筑能耗均有重要意义。本文基于遗传算法和 Grasshopper 平台搭建了一套基于太阳辐射利用的街区形态优化设计方法，以武汉市办公街区为研究对象，选取了 3 种典型街区形态，以冬季太阳辐射量与夏季太阳辐射量为优化目标对街区的形态进行优化。结果显示，优化后 3 种街区的冬夏季太阳辐射量差值均实现了提升，该方法为街区建筑太阳辐射被动式利用提供了一种新的思路。

关键词 太阳辐射；街区形态；夏热冬冷地区；遗传算法

Abstract Due to the particularity of climate in hot summer and cold winter areas, the energy consumption of refrigeration in summer and heating in winter is high. The utilization of solar radiation in buildings with hot summer and cold winter is of great significance for improving indoor thermal environment and reducing building energy consumption. Based on genetic algorithm and Grasshopper platform, this paper builds a set of block form optimization design method based on the utilization of solar radiation. Taking Wuhan office blocks as the research object, three typical block forms are selected to optimize the block form with the solar radiation amount in winter and the solar radiation amount in summer as the optimization objective. The results show that the difference values of solar radiation in winter and summer of the three kinds of blocks are improved after optimization, which provides a new idea for passive utilization of solar radiation in block buildings.

Keywords Solar radiation; Block form; Hot summer and cold winter areas; Genetic algorithm

徐燊，华中科技大学建筑与城市规划学院，教授，博士生导师。
侯俊豪，华中科技大学建筑与城市规划学院，硕士研究生。
谢孟举，华中科技大学建筑与城市规划学院，博士研究生。
桑梦城，华中科技大学建筑与城市规划学院，硕士研究生。

青岛里院冬季公共空间微气候实测分析

Qingdao Li House Winter Public Space Microclimate Measurements

邹雨菲，耿雪川，李津玮，赵文轩
Zou Yufei, Geng Xuechuan, Li Jinwei, Zhao Wenxuan

摘要 历史街区公共空间改造是存量时代下我国城市更新工作的重点之一，其舒适度是市民日常公共生活的本源诉求。为探究冬季公共空间不同下垫面特性与微气候环境和使用者感受之间的关系，本文结合微气候实测和热舒适问卷调查，对青岛“里院”公共空间进行微气候环境实地监测，比较分析不同绿化形式与硬质铺装条件下微气候因子和热舒适度的变化趋势。研究表明，绿化形式对白天空气温湿度影响较小，对风速、黑球温度影响较大。建筑物与植物的阴影为影响测点温度的主要原因，灌木与乔木对于风起到很大的减弱作用。

关键词 微气候；公共空间；历史街区；热舒适

Abstract The renovation of public spaces in historic districts is one of the key priorities of urban regeneration in China in the age of inventory, and the comfort of historic public spaces is a fundamental demand of people in their daily public life. In order to investigate the relationship between different substrate characteristics and microclimate environment and the users' perception of public spaces in winter, this study combined microclimate measurements and thermal comfort questionnaires to observe the microclimate environment of public spaces in Li House of Qingdao, and compared and analyzed the microclimate factors and thermal comfort under different greening forms and hard paving conditions. The study shows that greenery has less influence on daytime air temperature and humidity, and more influence on wind speed and black bulb temperature. The shadows of buildings and vegetation are the main cause of the temperature at the measurement point, and shrubs and trees play a large role in attenuating the wind.

Keywords Microclimate; Public space; Historic district; Thermal comfort

邹雨菲，青岛理工大学建筑与城乡规划学院，硕士研究生。
耿雪川，青岛理工大学建筑与城乡规划学院，副教授。
李津玮，青岛理工大学建筑与城乡规划学院，硕士研究生。
赵文轩，青岛理工大学建筑与城乡规划学院，硕士研究生。

立体绿化对微气候影响实测分析

Measurement and Analysis of the Influence of Vertical Greening on Microclimate

张海滨，陈舒，孙佳奇
Zhang Haibin, Chen Shu, Sun Jiaqi

摘要 为研究立体绿化对微气候的影响，在夏季高温天气对重庆地区的立体绿化建筑物进行微气候实测分析，比较了绿化朝向，距外墙距离等因素对绿墙降温增湿能力的影响。实验结果表明西向绿墙的降温能力最好，南向绿墙最差；西向绿墙可实现 4.95℃的冷却效果，南向绿墙只能实现 0.47℃的冷却效果；朝向对增湿能力的影响与对降温能力的影响相比较弱。立体绿化的影响范围较广，降温能力比较好，研究的结果有望证实街区尺度实施立体绿化可以改善城市微气候，为街区层面微立体绿化提供设计指导。

关键词 立体绿化；微气候；降温；增湿；朝向

Abstract In order to study the effect of vertical greening afforestation on microclimate, the micro-climate of vertical greening afforestation buildings in Chongqing was measured and analyzed in summer hot weather, and the effect of greening orientation, distance from external wall and other factors on the cooling and humidification ability of green wall was compared. The experimental results show that the west facing green wall has the best cooling ability, while the south facing green wall has the worst cooling ability. The western green wall can achieve a cooling effect of 4.95°C, while the southern green wall can only achieve a cooling effect of 0.47°C. The effect of orientation on humidification ability was not as significant as that of cooling. Vertical greening greening has a wide range of influence, and its cooling ability is relatively good. The results of this study are expected to confirm that the implementation of vertical greening greening at the block scale can improve the urban microclimate and provide design guidance for the micro-vertical greening greening at the block level.

Keywords Vertical greening; Microclimate; Cooling; Humidification; Orientatio

张海滨，重庆大学建筑城规学院，副教授，研究生导师，建筑技术科学系副系主任。
陈舒，重庆大学建筑城规学院，硕士研究生。
孙佳奇，重庆大学建筑城规学院，硕士研究生。

城市下垫面对街区热岛效应影响的模拟研究

Numerical Simulation on Influence of Urban Underlying Surface on Heat Island Effect of Blocks

杨子馨，傅秀章，谢华荣，李永辉 *
Yang Zixin, Fu Xiuzhang, Xie Huarong, Li Yonghui

杨子馨，东南大学建筑学院，硕士研究生。
傅秀章，东南大学建筑学院，副教授。
谢华荣，东南大学建筑学院，助理研究员。
李永辉，东南大学建筑学院，副教授，建筑技术研究所副所长。

摘要 本文基于 Rhino & Grasshopper 平台构建南京东南大学四牌楼校区的街区模型，通过 Dragonfly 插件模拟分析下垫面的反射率和绿地率对街区热岛效应的影响规律。结果表明：冬夏季街区热岛强度范围分别为 −4.4~7. 9℃和 −1.4~6.6℃；路面反射率和绿地率的增大有利于降低热岛强度，尤其对夏季的调节作用更为明显；提高路面反射率对白天热岛效应的缓解效果更强，而提高绿地率对夜间热岛效应的缓解效果更强。本研究可为南京城市街区规划设计提供理论指导和技术支持，以尽可能地缓解城市热岛效应。

关键词 热岛效应；路面反射率；绿地率；下垫面；Dragonfly 模拟

Abstract This study built an urban block model of Sipailou campus of Nanjing Southeast University based on Rhino & Grasshopper and used Dragonfly software to simulate and analyze the influence of pavement reflectivity and green space rate of urban underlying surface on the heat island effect of blocks. The results showed that: The range of heat island intensity in winter and summer was -4.4~7.9℃ and -1.4~6.6℃ respectively. The increase of pavement reflectivity and green space rate was beneficial to reduce the heat island intensity, especially in summer. Increasing the pavement reflectivity and the green space rate had a stronger mitigation effect on the daytime and the nighttime heat island effect, respectively. This study can provide theoretical guidance and technical support for urban block planning and design of Nanjing city to alleviate the urban heat island effect as much as possible.

Keywords Heat island effect; Pavement reflectivity; Green space rate; Underlying surface; Dragonfly simulation

---

反射涂料对夏热冬冷地区住宅小区微气候与负荷影响研究

A Study on the Effects of Reflective Coatings on Microclimate and Loads in Residential Communities in Hot Summer and Cold Winter Areas

余辉，张会波
Yu Hui, Zhang Huibo

余辉，上海交通大学设计学院，硕士研究生。
张会波，上海交通大学设计学院，副教授。

摘要 本研究针对住区建筑立面反射率对室外微气候和建筑冷、热负荷的影响进行量化和分析。基于 ENVI-met 和 EnergyPlus 模拟，通过正交实验法研究不同立面应用不同性能涂料对于人行高度处热舒适和建筑空调负荷的影响。结果表明，当所有立面应用低反射涂料时，人行高度 PET 达到最小值；提高涂料反射率对冬季空调负荷的影响高于夏季。当东立面应用高反射涂料，南、西立面应用低反射涂料，北立面采用中等反射的涂料时，全年空调负荷达到最小值。本研究结果可为夏热冬冷地区现存住宅小区的墙体改造提供参考。

关键词 反射涂料；微气候；空调负荷；正交实验

Abstract This study quantifies and analyzes the effect of building facade reflectance on outdoor microclimate and building cooling and thermal loads in settlements. Based on ENVI-met and EnergyPlus simulations, the effect of applying different performance coatings on different facades on thermal comfort and building air-conditioning load at pedestrian height is investigated by orthogonal experimental method. The results show that when low reflective coatings are applied to all facades, the pedestrian height PET reaches the minimum value; the effect of increasing coating reflectivity on air-conditioning load is higher in winter than in summer. The annual air-conditioning load reaches a minimum when high reflective coatings are applied to the east facade, low reflective coatings to the south and west facades, and medium reflective coatings to the north facade. The results can provide a reference for the wall retrofitting in existing residential communities in hot summer and cold winter areas.

Keywords Reflective Coating; Microclimate; Air-conditioning loads; Orthogonal experiments

城市水体对微气候影响的数值模拟研究进展

Research Progress in Numerical Simulation of the Impact of Urban Water Body on Microclimate

杨玉锦，李朝欣，张岂珲，赵敬源 *
Yang Yujin, Li Zhaoxin, Zhang Qihui, Zhao Jingyuan

杨玉锦，长安大学建筑学院，博士研究生，工程师。
李朝欣，长安大学建筑学院，博士研究生，讲师。
张岂珲，长安大学建筑学院，博士研究生。
赵敬源，长安大学，教授，博士生导师。

摘要 通常认为城市水体有助于改善局地微气候，但这种调节作用在不同时空尺度上的影响远比设想复杂，数值模拟是异质性城市环境微气候评估和优化设计策略的重要工具。本文综述了近年设计领域利用数值模拟方法开展的城市水体微气候影响研究，详细描述了已有工具方法的物理机制和计算模型，提出了数值模型优化原则，为研究人员提供了数值模拟的理论知识，对未来的城市微气候模型优化提出方向。

关键词 城市水体；数值计算；下垫面；物理模型

Abstract It is widely accepted that urban water bodies contribute to the improvement of local microclimates, but the effects of such modulation at various spatial and temporal scales are far more complex than anticipated. This article reviews recent work on the effects of urban water bodies on microclimate using numerical simulation methods in the design field, describes in detail the physical mechanisms and computational models of existing tool methods, proposes the principle of numerical model optimization, gives researchers theoretical knowledge of numerical simulations, and makes recommendations for the future optimization of urban microclimate models.

Keywords Urban water body; Numerical calculation; Underlying surface; Physical model

---

天井组合对漳州竹竿厝改造中的风环境影响模拟

Simulation of the Wind Environment Impact of Patio Combination on the Transformation of Zhugan Cuo in Zhangzhou

林姝颖，赖婉玲，吴正旺
Lin Shuying, Lai wanling, Wu Zhengwang

林姝颖，华侨大学建筑学院，硕士研究生。
赖婉玲，华侨大学建筑学院，硕士研究生。
吴正旺，华侨大学建筑学院，教授。

摘要 竹竿厝是闽南漳州的传统建筑，其特征之一是有多个天井，具有良好的通风和采光效果。在竹竿厝的重建中，其结构、材料及层数常常发生较大变化。采取何种天井组合最有利于自然通风，对于改善居住环境至关重要。基于当地气候，采用 Airpak 3.0.16 软件对其风环境进行模拟，分析天井组合的布局、形式及高度对风环境的影响，挖掘竹竿厝的风压、热压通风潜力，进一步改善竹竿厝的内部通风条件，有利于竹竿厝在侨乡振兴中的传承与发展。

关键词 竹竿厝；数值模拟；气候适应性；热压通风；风压通风

Abstract Zhugan Cuo is a traditional architecture in Zhangzhou, southern Fujian, characterized by multiple courtyards and excellent ventilation and lighting effects. In the reconstruction of Zhugan Cuo, there have been significant changes in its structure, materials, and number of floors. It is crucial to choose the combination of courtyards that is most conducive to natural ventilation in order to improve the living environment. Based on the local climate, Airpak 3.0.16 software was used to simulate the wind environment, analyze the impact of the form and height of the courtyard combination on the wind environment, explore the potential of wind pressure and thermal ventilation in Zhugan Cuo, and further improve the internal ventilation conditions of Zhugan Cuo, which is conducive to the inheritance and development of Zhugan Cuo in the revitalization of Qiaoxiang.

Keywords Zhugan Cuo; Numerical simulation; Climate adaptability; Hot press ventilation; Wind pressure ventilation

乔木遮挡时建筑外墙与邻近地物间的角系数计算方法研究

A Study on the Calculation Method of the View Factor between the Exterior Wall of a Building and Adjacent Features Under Arbor Shielding

张宽羽，赵晓悦，张泰龙，齐锋
Zhang Kuanyu,
Zhao Xiaoyue,
Zhang Tailong, Qi Feng

张宽羽，赵晓悦，浙江农林大学，硕士研究生。
张泰龙，浙江农林大学，博士研究生。
齐锋，浙江农林大学，教授，博士生导师。

摘要 乔木对邻近建筑外墙与周边墙面、地面等地物表面间的角系数及辐射换热过程都会产生较大影响。本文提出模拟光学投影法来计算外墙与邻近地物表面间角系数。首先计算无乔木条件下，墙面与邻近地物表面间的角系数。然后，用模拟软件计算存在树冠遮挡时的可视率，再计算此时的角系数。最后，用该角系数建立乔木遮挡下，墙面与其他表面间的长波辐射方程，以求解辐射理论值。分别对比墙面投入辐射和有效辐射的理论值与实测值，发现模拟光学投影法的误差率不超过3%。该方法可进一步研究乔木冠径、墙冠距及种植方位等因子对建筑外墙热辐射扰动的影响。

关键词 角系数；热辐射；球状树冠；模拟光学投影法

Abstract The view factors (VFs), between the exterior walls and other surfaces and the heat exchange will be affected if any canopy is close to the building. The simulated optical projection method is proposed to calculate the VFs from the exterior wall to other surfaces. Without any canopies, the VFs are first solved. Then, the simulation software is applied to obtain the visibility at differential elements with occlusion by the canopies and the VFs are acquired. Finally, the VFs are used to establish the radiation exchange equations and to calculate the theoretical values of long-wave radiation under the occlusion of the canopy. Comparing the theoretical and measured values of radiosity intensity and projective radiation at different points, it is found that the error rate of the simulated optical projection method is within 3%. The method can be used to further study the effects of factors such as the diameter of the canopy, wall-canopy distance, and orientation on the thermal radiation disturbance of the building.

Keywords View factor; Thermal radiation; Spherical canopy; Simulated optical projection method

城市街谷中行人的动态热经历

Study on the Dynamic Thermal Experience of Pedestrians in Urban Street Canyons

赵辉辉，许亘昱，石玉蓉，李晶磊，赵立华*，张宇峰
Zhao Huihui, Xu Genyu, Shi Yurong, Li Jinglei, Zhao Lihua, Zhang Yufeng

赵辉辉，华南理工大学建筑学院，博士研究生。
许亘昱，华南理工大学建筑学院，博士研究生。
石玉蓉，中山大学大气科学学院，博士后。
李晶磊，香港大学，博士研究生。
赵立华，华南理工大学建筑学院，教授。
张宇峰，华南理工大学建筑学院，教授。

摘要 本研究以广州中心城区三维地表模型为对象，采用路径分析、空间形态分析、日影分析和热辐射场模拟等方法，探究了城市街谷中行人受热辐射影响下的动态热经历特征。研究发现，夏季晴天日间，行人在阳光区和阴影区行走的平均时长与太阳高度角之间存在明显的指数函数关系，而行人路径上的日影交替数量、平均辐射温度突变量均和太阳高度角之间存在明显的二次函数关系。相较于针对户外特定空间的热舒适评价，城市行人的热舒适研究需要进一步考虑动态热经历对出行全程的热舒适影响。

关键词 行人热舒适；动态热经历；热辐射；出行路径；城市街谷

Abstract In this study, utilizing the three-dimensional surface model of the central urban area in Guangzhou, we employed methods such as path analysis, spatial morphology analysis, solar shading analysis, and thermal radiation field simulation to investigate the characteristics of pedestrians' dynamic thermal experience under the influence of thermal radiation in urban street canyons. The study revealed that during sunny summer days, there is a clear exponential relationship between the average duration of pedestrian movement in sunny and shaded areas and the solar altitude angle. Moreover, the number of alternating shadows along pedestrian paths and the average radiation temperature perturbation show a distinct quadratic relationship with the solar altitude angle. In comparison to the thermal comfort evaluation at specific locations, further consideration of the dynamic thermal experience's impact on pedestrians' entire journey is needed in urban outdoor thermal comfort research.

Keywords Pedestrian thermal comfort; Dynamic thermal experiences; Thermal radiation; Travel paths; Urban street canyons

基于文献计量法的水体热缓释研究演进趋势分析

Analysis of Cooling Effect of Water Body Researches' s Trends Based on Bibliometric

王佳婧，刘欣蕊，陈俊，
杨文健，王岩 *
Wang Jiajing, Liu Xinrui,
Chen Jun, Yang Wenjian,
Wang yan

摘要 为了掌握水体热缓释效应的研究现状和发展趋势，给相关课题的深入开展提供科学依据，本文采用文献计量法，使用 CiteSpace 软件对近 20 年中国知网数据库中的相关文献进行综合知识图谱分析。结果显示，我国水体热缓释效应的主要研究对象为景观格局和地表温度；多采用实地测试和数据模拟等方法；研究热点从下垫面类型、景观效应演进到蓝绿空间的影响效率和人体热舒适等方面；同时本领域的研究方向和国家政策引导紧密关联；未来针对微气候营造和室外人体热舒适等方面仍会是研究热点。

关键词 水体；热缓释效应；CiteSpace；知识图谱

Abstract In order to grasp the research status and development trend of cooling effect of water and to provide the scientific basis for in-depth research. In this paper, the Knowledge graph analysis method and CiteSpace software are used to analyze the relevant literatures in CNKI database in the past 20 years. The results show that the main research objects of the cooling effect of water are landscape pattern and surface temperature, and the field test and data simulation methods are used. The research focuses on the underlying surface type,landscape effect evolution to the influence efficiency of blue-green space. and human thermal comfort The research directions in this field is closely related to the national policies, and the research on microclimate construction and outdoor thermal comfort of human body will remain a hot topic in the future.

Keywords Water of body; Cooling effect; CiteSpace; Knowledge graph

王佳婧，刘欣蕊，陈俊，天津城建大学建筑学院，硕士研究生。
杨文健，天津城建大学国际工程学院，硕士研究生。
王岩，天津城建大学建筑学院，工学博士，教授。

透水混凝土地砖对中庭微气候调节效果的数值模拟分析

Numerical Analysis of the Microclimate Modifying Effect of Water-permeable Concrete Bricks in an Atrium

金华，王丽，何江
Jin Hua, Wang Li,
He Jiang

摘要 透水混凝土地砖作为改善中庭地面上方热环境的有效措施之一，它的降温持续效果受到太阳辐射影响极大。为探究饱和湿润透水混凝土地砖对中庭地面上方微气候的降温调节规律，采用数值模拟方法分析了湿润透水混凝土砖在中庭各区域的太阳辐射量与表面温度的变化规律以及对人体热舒适度的调节效果，并量化了湿润透水混凝土地砖在中庭地面各区域的降温时长。研究结果表明，在中庭两侧区域使用透水混凝土地砖，比中间区域使用具有更好的降温效果，对人体热舒适度的调节效果也会更高。

关键词 中庭地面；透水混凝土地砖；蒸发降温；微气候；数值模拟

Abstract Water-permeable concrete bricks are commonly used as an effective passive measure to control the surface temperature rise of the ground in an atrium. The cooling effect of the bricks is greatly influenced by solar radiation. In order to investigate the modifying effect of saturated permeable concrete bricks on the microclimate above the atrium ground under different solar radiation, this study used a numerical simulation approach to simulate surface temperature variation of wet permeable concrete bricks and thermal comfort indexes in different areas of the atrium ground. This paper also shows the cooling duration of the water-permeable concrete bricks in different areas of the atrium ground. The analysis results indicate that the water-permeable concrete bricks has a better cooling effect on the sides of the atrium than the middle area, and the modifying effect on human thermal comfort will also be greater.

Keywords Atrium ground; Water permeable; Concrete brick; Evaporative cooling; Microclimate; Numerical simulation

金华，广西大学土木建筑工程学院，硕士研究生。
王丽，广西大学土木工程学院建筑规划系，副教授。
何江，广西大学土木工程学院建筑规划系，教授。

第 7 篇

健康智慧的城市与建筑

基于细颗粒物分布的健康城市街区空间形态设计指标研究

Study on Spatial Form Design Index of Healthy Urban Blocks Based on Fine Particulate Matter Distribution

王薇，潘辉
Wang Wei, Pan Hui

摘要 研究通过移动监测与固定监测相结合的方法采集了 2022 年夏季合肥市芙蓉社区的 $PM_{2.5}$ 质量浓度数据，得到城市街区 $PM_{2.5}$ 空间分布特征，分析了街区空间形态设计指标与 $PM_{2.5}$ 质量浓度的相关性。结果表明：①街区细颗粒物质量浓度整体上呈现出由高到低（依次为施工区域、主路区域、道路交叉口、支路区域、绿化区域）的空间分布特征。②建筑高度比与细颗粒物质量浓度呈负相关，街谷高宽比、街谷长宽比与细颗粒物质量浓度呈正相关。

关键词 城市街区；健康城市；细颗粒物；空间形态；设计指标

Abstract By combining mobile monitoring with fixed monitoring, this study collected $PM_{2.5}$ mass concentration data of Furong Community in Hefei in the summer of 2022, obtained the spatial distribution characteristics of $PM_{2.5}$ in urban blocks, and analyzed the correlation between the spatial form design index of blocks and $PM_{2.5}$ mass concentration. The results showed that: ① On the whole, the mass concentration of fine particulate matter in the block showed the spatial distribution characteristics from high to low in order of construction area(main road area, road intersection, branch road area and green area). ② The building height ratio was negatively correlated with the mass concentration of fine particulate matter, while the height-width ratio of the street canyon and the length-width ratio of the street canyon were positively correlated with the mass concentration of fine particulate matter.

Keywords City block; Healthy cities; Fine particulate matter; Spatial form; Design indexes

王薇，合肥工业大学建筑与艺术学院，博士，教授。
潘辉，安徽建筑大学建筑与规划学院，硕士研究生。

基于 GIS 的南京主城区建成环境对空气质量的影响研究

Study on the Influence of Built Environment on Air Quality in Nanjing Main Urban Area Based on GIS Technology

李琴，彭昌海
Li Qin, Peng Changhai

摘要 城市空气质量与人们的健康息息相关。本文以南京主城区为例，从城市尺度定量探讨建成环境对城市空气污染的影响。首先对南京主城区的空气质量监测数据进行统计分析，探讨空气质量的空间分布现状特征。然后利用 ArcGIS 软件对主城区的空气质量进行空间分布模拟，并建立主城区空气质量与用地现状数据分析库。同时构建城市建成环境影响空气质量的指标体系，并通过 SPSS 软件进行相关性分析，得到南京主城区建成环境与空气质量之间的关系，探讨对空气质量具有显著影响的城市建成环境指标，为城市尺度的建成环境优化提供参考。

关键词 建成环境；空气质量；ArcGIS；影响机制；南京主城区

Abstract Urban air quality is closely related to people's health. Taking the main urban area of Nanjing as an example, this paper discusses the effect of the built environment on air quality from the urban scale. Firstly, this research analyzes the air quality in the main urban area and discusses the characteristics of spatial distribution. Secondly, ArcGIS is used to simulate the spatial distribution of air quality. Based on ArcGIS, this paper establishes the database of air quality and land use and constructs the index system of the impact of the urban built environment on air quality. Thirdly, the correlation analysis is carried out by SPSS to obtain the relationship between the built environment and air quality, so as to explore the indicators of urban built environment that have a significant impact on air quality, which provides reference for the optimization of urban built environment.

Keywords The built environment; Air quality; ArcGIS; Influence mechanism; Nanjing main urban area

李琴，东南大学建筑学院，硕士研究生。
彭昌海，东南大学建筑学院，教授。

成都市地下商业建筑空气污染物浓度及影响因素研究

Study of Air Pollutant Concentrations and Influencing Factors in Underground Commercial Buildings in Chengdu

李雨蒙，吴聃，张樱子，左丽君
Li Yumeng, Wu Dan, Zhang Yingzi, Zuo Lijun

摘要 为了调查成都市地下商场建筑空气污染情况，本文使用 PPM-HTV 甲醛检测仪、ppbRAE 手持式 VOCs 检测仪和 RAD7 测氡仪对成都市 7 个地下商业中的空气进行采样和监测。结果表明成都市地下商场中的甲醛和 TVOC 超标严重，超标率分别为 42.9% 和 85.7%，而所有地下商场的氡浓度均低于标准限值；大型地下商场的甲醛和 TVOC 超标率分别为 66.7% 和 100%，其污染状况比中小型商场更为严重；室外空气污染对地下商场空气污染物浓度贡献较小，而室内装修和商品售卖种类是影响地下商场污染物浓度的主要因素。

关键词 地下空间；空气质量；甲醛；TVOC；氡；影响因素

Abstract To investigate the building air pollution in underground shopping malls in Chengdu, this paper used a PPM-HTV formaldehyde detector, ppbRAE handheld VOCs detector, and RAD7 radon detector to sample and monitor the air in seven underground commercial buildings in Chengdu. The results showed that formaldehyde and TVOC in underground shopping malls in Chengdu exceeded the standard seriously, with exceedance rates of 42.9% and 85.7% respectively, while the radon concentrations in all underground shopping malls were below the standard limit; the exceedance rates of formaldehyde and TVOC in large underground shopping malls were 66.7% and 100% respectively, and their pollution conditions were more serious than those in small and medium-sized shopping malls; outdoor air pollution contributed less to the air pollutant concentrations in underground shopping malls Outdoor air pollution contributed less to the concentration of air pollutants in underground shopping malls, while the indoor decoration and the types of goods sold were the main factors affecting the concentration of pollutants in underground shopping malls.

Keywords Underground space; Indoor air quality; Formaldehyde; TVOC; Radon; Influencing factors

李雨蒙，西南交通大学建筑学院，硕士研究生。
吴聃，西南交通大学建筑学院，中级，讲师。
张樱子，西南交通大学建筑学院，正高级，学院助理。
左丽君，西南交通大学机械工程学院，硕士研究生。

健康建筑理论下的北方高校宿舍楼物理环境改造研究——以内蒙古工业大学宿舍楼为例

Research on the Physical Environment Transformation of Dormitory Buildings in Northern Colleges and Universities under the Theory of Healthy Architecture — Taking the Dormitory Building of Inner Mongolia University of Technology as an Example

白一帆，王崴 *
Bai Yifan，Wang Wei

摘要 宿舍楼作为大学生日常休憩、交流的重要场所之一，对人体舒适度及物理环境的要求尤为突出。本文首先基于健康建筑理论对建筑功能及建筑周边环境进行健康判断，找到影响人体舒适度和物理环境的主要因素；再借助内蒙古工业大学宿舍楼实际建筑区块进行改造策略模拟优化；最后利用健康建筑评判标准对新旧建筑模拟数据进行对比分析，着重从人体舒适度和物理环境角度得出建筑改造优化的可视性结论。

关键词 建筑改造；健康建筑；人体舒适度；物理环境；改造策略

Abstract As one of the important places for college students to rest and communicate in daily life, dormitory buildings have particularly prominent requirements for human comfort and physical environment. Firstly, based on the theory of healthy architecture, this paper makes a health judgment on the building function and the surrounding environment of the building and finds the main factors affecting human comfort and the physical environment. Then, with the help of the actual construction block of the dormitory building of Inner Mongolia University of Technology, the transformation strategy is simulated and optimized. Finally, the simulation data of new and old buildings are compared and analyzed by using the evaluation criteria of healthy buildings, and the visibility conclusion of building reconstruction optimization is obtained from the perspective of human comfort and the physical environment.

Keywords Building renovation; Healthy buildings; Human comfort; Physical environment; Transformation strategy

白一帆，内蒙古工业大学建筑学院，硕士研究生。
王崴，内蒙古工业大学建筑学院，副教授。

散热器表面温度对高校教室冬季热环境特性研究分析

Research and Analysis on the Influence of Surface Temperature of Radiators on the Thermal Environment Characteristics of University Classrooms in Winter

蒋婧，杨子毅，牛淑佳，狄育慧
Jiang Jing, Yang Ziyi, Niu Shujia, Di Yuhui

蒋婧，西安工程大学，副教授，硕士生导师。
杨子毅，牛淑佳，西安工程大学，硕士研究生。
狄育慧，西安工程大学，教授，硕士生导师。

摘要 为获取冬季高校教学建筑室内人员舒适的热环境，以西安市某高校典型教学建筑为研究对象，以最便捷改变温度的方式（即改变散热器辐射表面温度）进行了 3 种冬季工况热环境的数值模拟。通过模拟所得的教室内空气温度、PMV 和 PPD 云图分析出满足室内人员热舒适性的环境要求，耦合出西安市高校教学建筑冬季供暖状态下室内人体热舒适指标 PMV、PPD 模型公式，从而寻求能保证室内人员舒适的热环境状态参数。

关键词 教学建筑；热环境；数值模拟；平均热感觉；热舒适模型

Abstract In order to obtain a comfortable thermal environment forpeople inside university teaching buildings in winter, a typical teaching building in a certain university in Xi'an was taken as the research object, we conducted numerical simulations of the thermal environment under three winter operating conditions by changing the radiation surface temperature of the radiator, which is the most convenient way to change the temperature. Based on the simulated classroom air temperature, PMV and PPD nephogram, the environmental requirements that meet the thermal comfort of indoor personnel are analyzed, and then the PMV and PPD model formulas of indoor human thermal comfort indicators under the winter heating state of Xi'an University teaching buildings are coupled, so as to find the thermal environment state parameters that can ensure the comfort of personnel indoors.

Keywords Teaching building; Numerical simulation; Thermal comfort model; Mean thermal sensation; Adaptive predicted mean vote model

---

“绿色”涂料中的新污染物检测及暴露风险评估

Detection and Exposure Risk Assessment of New Chemicals in “Green” Paint Products

范钰杰，魏红岩，许瑛
Fan Yujie, Wei Hongyan, Xu Ying

范钰杰，清华大学建筑学院，博士研究生。
魏红岩，北新集团建材股份有限公司，工程师。
许瑛，清华大学建筑学院，副教授。

摘要 涂料是室内污染的一个重要来源。近年来，涂料行业正在逐步用低挥发性有机物和无挥发性有机物的水性涂料代替传统的溶剂性涂料。然而，为了达到好的产品性能，所谓的水性“绿色”涂料普遍大量使用了各类新型添加剂。我们分析了全球市场上 40 个不同品牌的水性涂料产品，首次发现中挥发性有机化合物（IVOCs）和半挥发性有机化合物（SVOCs）等一系列新型添加剂被广泛使用。我们选取涂料产品开展微舱实验，获得了新污染物从水性涂料散发的关键特性参数。通过结合污染物室内迁移模型和暴露风险评估，我们首次发现水性涂料中的新污染物可能会导致长期暴露风险。

关键词 水性涂料；新污染物；散发；暴露

Abstract Interior paint is an important source of indoor pollutants. Driven by regulatory requirements and market demands, the paint industry has replaced conventional “solvent-based” products with zero- or low- volatile organic compounds (VOCs) “water-based” paints. To achieve comparable performance and ensure product quality, water-based paints often contain various emerging additives. We analyzed 40 water-based paint products from different brands in the global market and found for the first time that a series of new additives, including intermediate volatile organic compounds (IVOCs) and semi-volatile organic compounds (SVOCs), were widely used. We have conducted micro-chamber experiments, and obtained key characteristic parameters for the emission of new chemicals from water-based paints. By combining indoor fate and transport model and exposure risk assessment, we have found that emerging chemicals in water-based paints may lead to long-term exposure risks.

Keywords Water-based paint; Emerging chemicals; Emission; Exposure

办公空间光热环境终端智能化的节能效果和舒适度影响实证研究

Empirical Research on the Energy-saving and Comfort of Intelligent Transformation of Office Built Environment

张淞博，苏航，陈星，尹璐，杨彪

Zhang Songbo, Su Hang, Chen Xing, Yin Lu, Yang Biao

摘要 现有的建筑设备智能控制系统可以有效降低建筑运行阶段的能耗，但其节能效果和对舒适度的影响，缺乏严谨的实证验证。本研究通过在实际办公空间中的物理环境终端智能化改造，同时在实验设计中引入了初始基线和平行对照组，开展了为期 129 天的“四阶段”实证实验，共有 69 名志愿者参与。实证期间收集了物理环境传感器、人员感知传感器和无线智能控制终端的数据，并分别在四个阶段对实验组和对照组开展基于双重差分法的结构化问卷调查。对比分析结果表明，该智能化改造能够有效降低 11.8% 用电量，同时未对舒适度造成负面影响。

关键词 建成环境；智能化改造；实证研究；双重差分法

Abstract Intelligent control systems can effectively reduce energy consumption during building operation, but there is a lack of rigorous empirical research of its energy-saving effects and impact on comfort. This study carried out a 129-day "four-stage" study by intelligently retrofitting the intelligent system in real office spaces, while introducing an initial baseline and a parallel control group into the experimental design. A total of 69 volunteers participated in the empirical experiment. During the demonstration period, the data of environment sensors, human perception sensors and wireless intelligent controllers were collected, using the Differences-in-Differences method (DID), structured questionnaire surveys were carried out for the treated group and the control group in four stages respectively. The comparative analysis results show that the intelligent transformation can effectively reduce lighting energy consumption by 11.8% without negatively affecting comfort.

Keywords Built environment; Intelligent transformation; Empirical research; Differences-in-Differences method

张淞博，苏航，哈尔滨工业大学（深圳）建筑学院，博士研究生。
陈星，深圳合一智控科技有限公司，清华大学工学硕士。
尹璐，哈尔滨工业大学（深圳）建筑学院，高级讲师。
杨彪，哈尔滨工业大学（深圳）建筑学院，副教授，博士生导师。

基于非线性规划的间歇式需求控制通风

Nonlinear Optimization-Based Intermittent Demand-Controlled Ventilation

牛顿，张胜

Niu Dun, Zhang Sheng

摘要 需求控制通风是一种高能效的通风控制方法。本课题组前期研究提出了间歇式需求控制通风，适用于定风量系统。本文进一步提出了一种基于非线性规划的间歇式需求控制通风方法，考虑室内空气质量和室外空气状态，规划风机的启停状态，保证室内空气质量的同时最小化新风能耗。针对我国不同气候区，与传统方法和现有的间歇式需求控制通风方法相比，新方法可以实现降低 18.6%~19.6% 和 5.5%~6.7% 的新风能耗。本文所提出的新方法有效降低了定风量系统的新风供应能耗，有助于开发低碳和健康的建筑环境。

关键词 需求控制；定风量系统；非线性规划；室内空气质量

Abstract Demand-controlled ventilation is an effective means of controlling ventilation while conserving energy. In previous research, our group proposed an intermittent demand control ventilation method that is compatible with constant air volume systems. In this paper, we present a new approach to intermittent demand control ventilation based on nonlinear programming. Our method takes into account both indoor air quality and outdoor air temperature/enthalpy, optimizing fan start and stop status to minimize energy consumption while ensuring indoor air quality. Compared with the traditional method and the existing intermittent demand control ventilation method, the new method can achieve a reduction of energy consumption of fresh air of 18.6%~19.6% and 5.5%~6.7% for different climate zones in China. Our new method contributes to the development of low-carbon and healthy building environments by effectively reducing energy consumption in the constant air volume system.

Keywords Demand control; Constant air volume system; Nonlinear optimization; Indoor air quality

牛顿，西安交通大学，硕士研究生。
张胜，西安交通大学，特聘研究员。

夏季二氧化碳浓度对学生学习效率的影响研究

Effect of Carbon Dioxide Concentration on Learning Efficiency in Summer

蒋婧，孙已明，秦石磊
Jiang Jing, Sun Yiming, Qin Shilei

摘要 教室作为中学生学习的主要场所，学习效率与其内部环境因素密切相关。本文采取主观问卷调查与学习效率测试实验相结合的方法，研究了夏季教室舒适温度 24℃无风状态下 CO_2 浓度对学习效率的影响。在模拟教室环境中设置了空气流速 0 m/s 的条件下，CO_2 分别为 600 ppm、1200 ppm、1800 ppm 和 2400 ppm 的四组实验工况，分析 CO_2 浓度对受试者学习效率影响关系。结果表明：CO_2 浓度增加会降低环境舒适感和加重病态建筑综合征症状；学习效率随 CO_2 浓度的增加而下降；不同 CO_2 浓度对学习效率对正确率、反应时间和学习效率的影响程度不同。

关键词 二氧化碳浓度；学习效率；主观评价；实验研究

Abstract Classroom is the main place for middle school students to learn, and learning efficiency is closely related to its internal environmental factors. This paper studies the effect of CO_2 concentration on learning efficiency under the comfortable temperature of 24℃ in summer without wind by combining subjective questionnaire and learning efficiency test experiment. Under the condition of air flow rate of 0 m/s in the simulated classroom environment, four groups of experimental conditions with CO_2 of 600 ppm, 1200 ppm, 1800 ppm and 2400 ppm were set to analyze the relationship between CO_2 concentration and learning efficiency of subjects. The results show that the increase of CO_2 concentration will reduce the environmental comfort and aggravate the symptoms of sick building syndrome; The learning efficiency decreases with the increase of CO_2 concentration; Different CO_2 concentrations have different effects on learning efficiency, correct rate, reaction time and learning efficiency.

Keywords Carbon dioxide concentration; Learning efficiency; Subjective evaluation; Experimental study

蒋婧，西安工程大学，副教授，硕士研究生导师。
孙已明，西安工程大学，硕士研究生。
秦石磊，西安工程大学，硕士研究生。

基于街景地图的长时序历史街区空间品质研究

Spatial Quality Measurement of Long Time Series Historic Districts Based on Street View Maps

王凤琳，王雨秋，张弘驰 *
Wang Fenglin, Wang Yuqiu, Zhang Hongchi

摘要 历史街区是我国城市更新的主要对象，对环境品质要求较高，但在城市化进程中普遍遭到“建设性破坏”。本文以大连东关街历史文化街区为例，运用 deeplab v3+ 语义分割对街景图片绿视率、街道开敞度和界面围合度等客观指标进行量化分析，并进行长时序纵向比较，以此评估历史街区发展与保护的动态变化与问题。研究发现东关街存在绿化面积逐渐减少、绿化覆盖率整体偏低、街区围合度逐年增加、开敞度呈下降趋势等问题并对其原因进行分析。本研究将为历史街区保护规划与空间设计提供参考。

关键词 历史街区；空间品质；街景图片；长时序

Abstract Historic districts are the main objects of urban renewal in China, which require high environmental quality but are generally "constructively destroyed" in the process of urbanization. The study took the Dongguan Street historical and cultural district of Dalian as an example, and used deeplab v3+ semantic segmentation to quantify the objective indicators such as green view rate, street openness and interface enclosure of streetscape pictures, and made a long time series longitudinal comparison, so as to assess the dynamic changes and problems of the development and conservation of the historical district. The study found that the green area of Dongguan Street is gradually decreasing, the overall green coverage rate is low, the enclosure of the street is increasing year by year, and the openness of the street is decreasing, and analyzed the causes of the problems. The study will provide reference for the conservation planning and spatial design of the historic districts.

Keywords Historic districts; Spatial quality; Street view pictures; Long time series

王凤琳，王雨秋，大连理工大学建筑与艺术学院，硕士研究生。
张弘驰，大连理工大学建筑与艺术学院，博士，讲师。

居住者行为与室内空气环境关联性研究：以某小户型住宅为例

The Correlation Study Between Occupant Behavior and Indoor Air Environment: Taking a Small Apartment as an Example

余翔宇，宋德萱，刘益
Yu Xiangyu,
Song Dexuan, Liu Yi

摘要 在高密度的居住环境下，由于室内活动频繁、空间布局紧凑，居住者的行为变化极易对室内空气环境舒适度造成影响。本文以上海市某小户型空间为研究对象，通过仪器不间断测试室内多个点位的 CO_2 浓度、温度和相对湿度等空气环境参数，科学呈现长期居住环境下居室内的空气环境质量，并通过空气环境数据和日常行为的关联性分析，揭示居住者日常行为对室内空气质量的影响及其变化特征。最后从建筑师的视角，提出提升空间环境品质的改进设计策略，营造健康的室内空间环境。

关键词 室内空气环境；小户型；居住者行为；关联性；改进策略

Abstract In the high-density residential environment, due to frequent indoor activities and compact space layout, the change of occupant behavior is very easy to affect the comfort of the indoor air environment. The paper takes a small Apartment in Shanghai as the research object, and tests the indoor air environment parameters such as CO_2 concentration, air temperature and relative humidity at multiple points for a continuously, scientifically presents the air environment quality under long-term residential environment, and reveals the impact of occupant behavior on indoor air quality and its change characteristics through the correlation analysis of air environment data and daily behavior. Finally, from the perspective of architects, the paper puts forward design strategies to improve the quality of space environment, to create a healthy indoor space environment.

Keywords Indoor air environment; Small apartment; Occupant behavior; Correlation; Improvement strategy

余翔宇，同济大学建筑与城市规划学院，博士研究生。
宋德萱，同济大学建筑与城市规划学院，教授，博士生导师。
刘益，同济大学建筑与城市规划学院，博士研究生。

智慧城市理念下历史文化名村保护中的信息技术手段探索

Research on the Means of Information Technology in the Protection of Historical and Cultural Villages Under the Concept of Smart City

姚凌涵
Yao Linghan

摘要 在智慧城市理念快速发展的背景下，历史文化名村保护作为城市建设与管理的重要组成部分，需要利用新技术来解决旧问题。本文以历史文化名村的保护流程为线索，从保护规划编制前的数据收集，到保护规划编制中的数据处理与决策制定，再到保护规划编制后的日常管控，概述了各阶段的信息技术手段应用，分析其优劣势。基于上述分析，探究信息技术在历史文化名村保护中的全流程应用，利用 GIS 与 BIM 搭建几何模型、非物理数据信息模型、参数模型三位一体的融合模型，探索历史文化名村保护、管理与更新的优化方法。

关键词 智慧城市；历史文化名村；保护规划；现状调查

Abstract In the context of the rapid development of the concept of smart city, the protection of historical and cultural villages, as an important part of urban construction and management, needs to use new technologies to solve old problems. Taking the protection process of historical and cultural villages as a clue, this paper summarizes the application of information technology means in each stage, and analyzes its advantages and disadvantages, from data collection before the preparation of protection planning, data processing and decision-making in the preparation of protection planning, to daily control after the preparation of protection planning. Based on the above analysis, this paper aims to explore the whole process application of information technology in the protection of historical and cultural villages, and build a trinity integration model of geometric model, non-physical data information model and parameter model based on GIS and BIM technology, in order to explore the optimization method of protection, management and update of historical and cultural villages.

Keywords Smart city; Historical and cultural village; Conservation planning; Current situation investigation

姚凌涵，华南理工大学，硕士研究生。

医院标识导视系统设计基本方法

The Basic Design Method of Hospital Signage Guidance System

焦瑞，乌云巴根
Jiao Rui, Wuyun Bagen

摘要 新冠疫情带给医疗建筑新的冲击，其中为避免寻路问题引起的交叉感染，医院标识导视系统的设计不容忽视。本文通过对近些年医院标识导视系统优化的相关文献阅读与对优秀设计案例的解析，从医院标识导视系统的基本概念、主要内容、设计原则，就其设计布局、造型、界面等方面，阐述了医院标识导视系统设计的基本方法并指出其不足，同时提出相关改进建议。

关键词 导视系统；标识；医院；设计方法

Abstract The COVID-19 has brought new implications to medical buildings. In order to avoid cross-infection caused by way-finding, the design of hospital signage system cannot be ignored. By analyzing the optimization of hospital signage system and relevant literature of excellent design cases in recent years, this paper expounds the basic method of hospital logo guide system design, points out its shortcomings, and puts forward relevant suggestions for improvement,from the basic concept, main content, design principle, design layout, shape and interface.

Keywords Guide system; Signage; Hospital; Design method

焦瑞，华侨大学建筑学院，硕士研究生。
乌云巴根，华侨大学建筑学院，讲师。

冬季外墙热工性能波动对室内健康物理环境的影响

Iinfluence of External Wall Thermal Performance Variation on Healthy Indoor Environment in Winter

王国乾，马茹婷，刘大龙
Wang Guoqian, Ma Ruting，Liu Dalong

摘要 建筑外墙的热工性能波动在冬季尤为明显，室内健康物理环境是人体健康的建筑保障。本文以位于夏热冬冷气候区的江苏省淮安市居住建筑为研究对象，结合现行的热工设计规范和当地气候特征，通过对淮安周边地区采暖度日数的聚类分析，并采用三种外墙热工设计方案对比研究。本文提出装配式建筑外墙的保温宜采用免拆模保温板与淮安市当地居住建筑外墙的传热系数设计限值宜由 0.80W/(m^2·K) 提高到 0.50W/(m^2·K) 的热工策略，为当地改善冬季室内健康物理环境，提供了设计指引。

关键词 夏热冬冷气候区；冬季外墙；热工性能波动；健康物理环境；营造策略

Abstract The variation of the thermal performance of the external wall is particularly noticeable in winter, the healthy physical environment of the building space is the architectural guarantee of human health. This paper takes the residential buildings in Huai'an city, Jiangsu Province in hot summer and cold winter climate area as the research object, combined with the current thermal design specifications and local climate characteristics, through the cluster analysis of the surrounding areas of Huai'an, and uses three kinds of external wall thermal design schemes. It is proposed that the thermal engineering design limit of the heat transfer coefficient of the exterior wall of prefabricated buildings and the exterior wall of local residential buildings in Huai'an city should be increased from 0.80W / (m^2·K) to 0.50W / (m^2·K). It provides design guidelines to improve the local indoor healthy physical environment in winter.

Keywords Hot summer and cold winter climate zone; Winter exterior wall; Thermal performance fluctuation; Health physical environment; Construction strategy

王国乾，西安建筑科技大学建筑学院，工程师，博士研究生。
马茹婷，西安建筑科技大学建筑学院，博士研究生。
刘大龙，西安建筑科技大学建筑学院，副教授，博士生导师。

性能驱动下建筑初期方案生成
——基于 EvoMass 平台的建筑体量寻优

Performance-driven Generation in Early-stage Architectural Design — Building Massing Optimization Based on EvoMass

蔡兆聪，冉茂宇
Cai Zhaocong, Ran Maoyu

摘要 建筑性能寻优算法及其参数化生成已在建筑方案设计初期得到了较为广泛的运用。然而，传统寻优算法仅仅是对建筑某项性能的优化，其优化结果的单一性与建筑设计初期方案的多样性和发散性存在矛盾。本文采用 Grasshopper 环境中的 EvoMass 与 Ladybug Tools 系列的插件，首先构建基于性能优化的建筑参数化设计流程，然后以厦门地区某公共建筑设计为例，具体说明了其在建筑初步设计中的实施过程，以验证其可行性，可为建筑师在初设方案阶段提供参考。结果表明，该流程可对建筑体量进行不同方式的参数化生成，并且可以基于多目标和多结果进行建筑体量优化。

关键词 建筑性能优化；参数化设计；绿色建筑；遗传算法；EvoMass

Abstract Algorithmic optimization based on building performance and parametric generation of buildings have been widely used as the initial design methods for building solutions. However, the traditional optimization algorithm only optimizes a certain performance value of the building, and the singularity of the optimization result contradicts with the diversity and dispersion of the initial architectural design scheme. In this paper, we use EvoMass and Ladybug Tools series plug-ins in Grasshopper environment to build a parametric design process based on performance optimization of buildings. To verify its feasibility, this paper takes a public building design in Xiamen area as an example to illustrate how it can provide architects with solutions and performance references in the early design stage. The results show that the process can generate building massing parametrically in different ways and optimize it based on multiple objectives and results.

Keywords Building performance optimization; Parametric design; Green building; Genetic algorithm; EvoMass

蔡兆聪，华侨大学建筑学院，厦门市生态建筑营造重点实验室，硕士研究生。
冉茂宇，华侨大学建筑学院，厦门市生态建筑营造重点实验室，博士，教授。

基于机器学习的活动百叶遮阳控制策略与算法比较
——以夏热冬暖地区为例

A Movable Blind Control Strategy Based on Machine Learning and Algorithms Comparison —Taking a Hot Summer and Warm Winter Region as an Example

王振宇，孟庆林，汪俊松
Wang Zhenyu, Meng Qinglin, Wang Junsong

摘要 可根据外界环境情况调控的建筑动态遮阳有着更大的节能潜力和室内光热舒适度，其中控制策略是主导动态遮阳效果的核心因素之一。本文以总能耗最优为目标，探究了一种基于机器学习的遮阳控制策略，并对机器学习中四种算法（支持向量机、逻辑回归、决策树、K 临近算法）训练得出的遮阳控制策略的节能表现进行了比较。结果表明：通过机器学习的遮阳控制策略与学习集相比有较高的预测准确度，并在预测集上有很好的节能表现。其中 SVM 算法的表现最好，节能率最高。

关键词 遮阳控制；机器学习；建筑能耗；百叶遮阳

Abstract Dynamic shading of buildings that can be regulated according to the external environment has greater energy saving potential and indoor light and thermal comfort, where the control strategy is one of the core factors that dominates the effect of dynamic shading. In this study, a machine learning-based shading control strategy is explored with the goal of optimal total energy consumption. And the energy saving performance of the shading control strategies derived from four algorithms (support vector machine, logistic regression, decision tree, and K-nearestneighbor) are compared. The shading control strategy by machine learning has a high prediction accuracy compared to the learning set and has a good energy saving performance on the prediction set. The SVM algorithm had the highest performance with the highest energy savings.

Keywords Shading control; Machine learning; Building energy consumption; Venetian blind shading

王振宇，华南理工大学，硕士研究生。
孟庆林，华南理工大学，教授，博士生导师，中国建筑学会物理分会理事长。
汪俊松，华南理工大学，博士后。

基于多方利益的住宅品质分布模拟研究

Simulation of Housing Distribution Based on Multi-interest

任佳雯，周欣，石邢
Ren Jiawen, Zhou Xin, Shi Xing

摘要 提高城市居民的生活品质是任何城市建设的重要目标之一。因此，如何在有限的资源下尽可能地提高城市中的住宅品质，并合理规划城市中不同品质住宅的分布，具有复杂性且值得深入研究。本研究选取中国山东内陆快速城市化的典型区域山东省冠县作为研究区，充分考虑了城市土地利用特性以及城市内部多智能体的行为。以研究区域 2030 年城市发展功能分布格局为基础，对居民、开发商和政府三类智能体的行为进行定量化描述，并在此基础上设计模型迭代流程，构建基于多方利益的住宅品质模型。该模型能够对城市未来居住用地的品质进行预测，进而为后续进行土地利用情景分析提供模型基础。

关键词 多智能体模型；住宅品质；土地利用模拟

Abstract Improving the quality of life of city residents is one of the important goals of any city construction. Therefore, how to improve the quality of urban housing as much as possible and rationally plan the distribution of different quality of urban housing under limited resources is complicated and worthy of further study. In this study, Guanxian County of Shandong Province, a typical area of rapid inland urbanization in Shandong Province, China, is selected as the research area, and the characteristics of urban land use and the behavior of multi-agents within the city are fully considered. Based on the distribution pattern of urban development functions in the study area in 2030, a multi-interests-based housing quality model is constructed. The behavior of three types of intelligent agents, namely residents, developers and government, is quantitatively described, and the model iteration process is designed on this basis, which can predict the quality of urban residential land in the future, and thus provide a model basis for the subsequent land use scenario analysis.

Keywords Agent-based model; Housing quality; Land use simulation

任佳雯，同济大学建筑与城市规划学院，博士研究生。
周 欣，东南大学建筑学院，教授，博士生导师。
石邢，同济大学建筑与城市规划学院，教授，博士生导师。

基于 CFD 模拟的多层住区呼吸道传染病防控策略研究

Research on Prevention and Control Strategy of Respiratory Infectious Diseases in Multi-storey Residential Area Based on CFD Simulation

张涵琪，蒋清睿 *，叶海，罗志刚，涂慧君
Zhang Hanqi, Jiang Qingrui, Ye Hai, Luo Zhigang, Tu Huijun

摘要 2020 年，上海出台《关于完善重大疫情防控体制机制、健全公共卫生应急管理体系的若干意见》，其中强调了公共卫生防控体系建设的急迫性和重要性。社区作为人们生产生活的重要场景，建设公共卫生防控体系也成为重要的一环。本研究针对社区疫情风险之一的远程空中传播现象，以上海城市空间中最常见的多层住区为研究对象，基于 CFD 的污染物扩散模型模拟结果总结了社区不同风场条件下病毒的扩散特点，并基于模拟结果提出了应对病毒传染的公共卫生防控策略。

关键词 CFD 模拟；呼吸道传染病；公共卫生防控策略

Abstract In 2020, Shanghai issued *the Opinions on Improving the System and Mechanism of Major Epidemic Prevention and Control and Improving the Public Health Emergency Management System*, which stressed the urgency and importance of building a public health prevention and control system. As an important scene of people's production and life, the construction of community public health prevention and control system has become an important step. Focusing on the phenomenon of long-distance aerial transmission, one of the epidemic risks in communities, the study took the most common multi-storey residential area in Shanghai urban space as the research object. The simulation results of pollutant diffusion model based on CFD summarized the characteristics of virus diffusion under different wind conditions in communities, and proposed public health prevention and control strategies to deal with virus infection based on the results.

Keywords CFD simulation; Respiratory infectious disease; Public health prevention and control strategies

张涵琪，同济大学建筑与城市规划学院，硕士研究生。
蒋清睿，同济大学建筑与城市规划学院，博士研究生。
叶海，同济大学建筑与城市规划学院，副教授。
罗志刚，上海同济城市规划设计研究院，院长助理。
涂慧君，同济大学建筑与城市规划学院，教授。

基于感知联动的老年设施单元计算性设计与控制

Computational Design and Indoor-adjustment for the Module of Elder Facility Responding to Occupant Perception

何文晶，詹进，尹涤非
He Wenjing, Zhan Jin, Yin Difei

摘要 近年来，通过建筑优化设计与室内环境动态调控策略提升建筑性能成为了研究热点。本文探索和构建了老年设施单元计算性设计和基于感知联动的室内环境控制方法。针对寒冷地区老年设施的使用特点，通过计算性方法，完成实验单元的建筑设计；通过对老年人的生理和脑电数据采集，并结合问卷调查，确定室内舒适的温度区间；基于多重感知联觉需求，自控多性能集成式光热光伏界面系统，确保室内环境的达成。本文通过多源协同优化算法和智能控制的调节方法，来满足老年人的使用要求，实现健康、舒适与节能三重目标。

关键词 寒冷地区；老年设施；光电光热界面；感知联动

Abstract Recently, there is a rising interest in building optimization design and dynamic indoor-adjustment strategies aimed at building performance improvement. This work explored and established a holistic methodology for computational design for a module of an elder facility. This work also brought up an indoor-adjustment method in accordance with the occupant's perception. The design for the experimental module was completed based on data and building behaviors in cold regions. The appropriate indoor temperature range was determined through ergonomic data as well as a set of subjective questionnaires. The thermal comfort indoors was maintained by individual satisfaction as well as the self-controlled photothermal-photovoltaic interfaces. This work satisfied the elder's health, comfort, and energy savings demands by relying on computational design and indoor adjustment.

Keywords Cold regions; Facilities for the elderly; Photothermal-photovoltaic interfaces; Occupant perception

何文晶，山东建筑大学，教授，硕士生导师。
詹进，山东建筑大学，硕士研究生。
尹涤非，山东建筑大学，硕士研究生。

通风窗形状及位置对体育馆风压通风空气品质的影响

Investigation of Opening Position and Shape on Natural Cross Ventilation in Gymnasia

程征，许程骞，谢芳灵，张培峰 *
Zheng Cheng, Xu Chengqian, Xie Fangling, Zhang Peifeng

摘要 本文通过调整通风窗排列方式、窗台高度、水平间距 3 个几何参数，以换气次数、空气龄为自然通风空气品质评价指标，为体育馆运行过程中自然通风窗的调节应变提供建议。研究发现改变通风窗形态对体育馆进风总量影响较小，但是核心使用空间的空气龄分布变化显著：竖向排列开窗使看台及场地区平均空气龄明显下降；降低窗台高度使气流沿观众看台向下流动，显著改善体育馆场地区空气品质；加大通风窗水平间距有助于降低上风向看台及两侧看台空气龄。

关键词 自然通风；体育馆；通风窗形状；通风窗位置；空气龄

Abstract Three geometric parameters are evaluated, opening shapes, window sill heights and horizontal spacing, for cross natural ventilation in gymnasia using air age as performance indicators. It is found that opening configurations have small impacts on air exchange rate but significant influence on air age in core spaces in gymnasia. Vertical composite windows commit to reduce the average air age in all core spaces. Reducing opening sill heights can induce airflow move downward. Air age of the upwind spectator stands and side stands could be reduced by expanding the horizontal opening spacing.

Keywords Natural ventilation; Gymnasia; Opening shapes; Opening positions; Air age

程征，中国石油大学（华东）储运与建筑工程学院建筑系，副教授。
许程骞，谢芳灵，中国石油大学（华东）储运与建筑工程学院建筑系，本科生。
张培峰，中国石油大学（华东）储运与建筑工程学院建筑系，高级实验师。

晋北地区不同采暖方式农宅室内空气品质研究

Study on Indoor Air Quality of Rural Houses with Different Heating Methods in Northern Shanxi Province

张孟莹，董旭娟，冯晶
Zhang Mengying, Dong Xujuan, Feng Jing

张孟莹，太原理工大学建筑学院，硕士研究生。
董旭娟，太原理工大学建筑学院，副教授。
冯晶，太原理工大学建筑学院，硕士研究生。

摘要 本文通过对晋北地区农宅的采暖方式与室内空气品质的主观调研，对其中主要的 4 种采暖方式——空气能热泵暖气片式、燃煤锅炉暖气片式、燃煤锅炉地暖式和火炉火炕式的农宅室内主要污染物 $PM_{2.5}$、PM_{10}、TVOC 和 CO_2 的浓度进行 24 h 连续监测与对比分析。研究结果表明：晋北地区农宅 4 种采暖方式室内 TVOC 的浓度均超标较多；火炉火炕式室内 $PM_{2.5}$ 和 PM_{10} 的 24 h 浓度呈现间歇性高峰波动；室内空气品质从优到劣依次是空气能热泵暖气片式、燃煤锅炉地暖式、燃煤锅炉暖气片式和火炉火炕式。

关键词 室内空气品质；农宅；采暖方式；晋北地区

Abstract This paper conducted a subjective questionnaire survey on the heating methods and indoor air quality of rural houses in northern Shanxi Province. The indoor concentrations of $PM_{2.5}$, PM_{10}, TVOC and CO_2 of the four main heating methods were monitored continuously for 24 hours and analyzed. The results showed that the indoor TVOC concentrations of the four main heating methods all exceeded the standard by a large margin in northern Shanxi Province; the 24-hour concentrations of $PM_{2.5}$ and PM_{10} of Chinese stove and kang heating showed intermittent peak fluctuations; the heating methods corresponding to the indoor air quality from best to worst were air-source heat pump radiator heating, coal-fired boiler radiant floor heating, coal-fired boiler radiator heating and Chinese stove and kang heating.

Keywords Indoor air quality; Rural houses; Heating methods; Northern Shanxi Province

---

老年人热舒适研究热点及趋势可视化分析

The Hotspots and Trend Visualization Analysis of the Elderly Thermal Comfort Research

陈俊，刘欣蕊，王佳婧，杨文建，王岩 *
Chen Jun, Liu Xinrui, Wang Jiajing, Yang Wenjian, Wang Yan

陈俊，刘欣蕊，王佳婧，杨文建，天津城建大学建筑学院，硕士研究生。
王岩，天津城建大学建筑学院，工学博士，教授。

摘要 随着我国人口老龄化程度加剧，养老问题备受社会关注，党的二十大提出积极应对老龄化战略，因此，开展老年宜居研究意义深远。本文基于中国知网（CNKI）数据库中近 20 年的相关文献，运用文献计量软件（CiteSpace）对老年人热舒适研究的发文量、关键词共现和聚类、作者及合作网络等方面进行可视化分析。结果显示，本领域的研究经历了初始期、增长期和平稳期 3 个阶段；形成若干核心研究团队，但团队间的合作较少；"优化策略""现场实测""宜居改造"等仍是研究热点。

关键词 文献计量软件；可视化分析；老年人热舒适；热环境

Abstract As China's population is aging, the issue of elderly care is of great concern to society, and the 20th Party Congress put forward the strategy of actively coping with aging; therefore, it is of far-reaching significance to conduct research on elderly livability. Based on the relevant literature in the database of China Knowledge Network (CNKI) in the past two decades, this paper uses visual bibliometric software (CiteSpace) to visualize and analyze the number of publications, keyword co-occurrence and clustering, authorship and cooperation networks of thermal comfort research on the elderly. The results show that the research in this field has gone through three stages: initial, growth and stable; several core research teams have been formed, but there is less cooperation among teams; the research hotspots are still "optimization strategy", "field measurement" and "livable renovation".

Keywords CiteSpace; Visual analysis; Thermal comfort for the elderly; Thermal environment

湿热地区高校教室开窗位置对室内风环境的影响

Impact of Window Position on the Indoor Wind Environment of University Lecture Rooms in Hot and Humid Region

夏燕，邓寄豫 *
Yan Xia, Deng Jiyu

摘要 本研究针对夏热冬暖地区高校教室常见的推拉窗和悬窗，深入分析了其位置变化对室内风速分布及自然通风效果的影响。研究结果表明，对推拉窗而言，当来流风垂直吹向窗面时，将进口窗分散设置，可实现较好的室内气流分布；而当来流风与窗面斜交时，应尽量将进口窗靠近来流的上风向设置，同时尽量将出口窗靠近来流的下风向设置。而对悬窗而言，来流风与窗面垂直时，上悬窗（外开）洞口的位置最好设置在墙体中部，以确保气流覆盖室内大部分区域，下悬窗（外开）入口则尽量设置高些，以减缓风能衰减过程。

关键词 室内风环境；自然通风；开窗位置；高校教室；立面设计

Abstract Aiming at the sliding and hanging windows commonly used in university lecture rooms in hot summer and warm winter regions, this study explores the impact of window position on indoor wind velocity distribution and natural ventilation performance. The results indicate that for the sliding windows, the inlet openings should be scattered to achieve better indoor airflow distribution when the incoming airflow is perpendicularly flowing to the window. Otherwise, it should be set on the upwind side, while the outlet openings should be on the downwind side when the incoming wind is obliquely flowing to the window. Furthermore, the top-hung windows (outward-opening) are suggested to be positioned at the middle height of the wall to ensure that the airflow could cover most of the indoor area. The bottom-hung windows (outward-opening) are preferably set at a higher position to slow down the wind energy attenuation.

Keywords Indoor wind environment; Natural ventilation; Opening position; University lecture room; Façade design

夏燕，广东工业大学建筑与城市规划学院，硕士研究生。
邓寄豫，广东工业大学建筑与城市规划学院，讲师，硕士生导师。

第 8 篇

低碳城市与建筑

系统动力学模式下的建筑系统全生命周期可持续性研究

Investigation on the Whole Life Cycle Sustainability of Building System Under the Exploration of System Dynamics Methodology

张军学，张何，李劭雄，王荷池，吴国栋
Zhang Junxue, Zhang He, Li Shaoxiong, Wang Hechi, Wu Guodong

张军学，江苏科技大学建筑学系，博士，建筑学系系主任。
张何，佛罗里达大学设计、建筑与规划学院，博士，研究员。
李劭雄，上海市城市建设设计研究总院（集团）有限公司，工程师。
王荷池，湖北工业大学土木建筑与环境学院，博士，讲师。
吴国栋，苏州大学建筑学院，博士，高级工程师。

摘要 在碳中和及全球气候变化的背景下，建筑系统的生态可持续性研究变得热门。结合系统动力学模型和全生命周期的建筑系统分析，以实际建筑案例为落脚点，本研究对目标建筑系统的系统动力学模型进行了设计，特别是在反馈子系统的集成下，对比分析了优化前后模型的效率。研究结果揭示了建筑系统的动力学模型改进路径，并交叉能值计算，分析前后模型的可持续参数变化，验证反馈优化的效果，为系统动力学理念模型与建筑系统的深度耦合提供了参考方向。

关键词 建筑系统；系统动力学；全生命周期；能值可持续性

Abstract Under the background of carbon neutrality and global climate change, the ecological sustainability study of building systems becomes increasingly a hot topic. Combined with the system dynamics model and the whole life cycle method of the building system analysis, taking the actual building case as the starting point, the system dynamics model of the target building system is designed, especially under the integration of the feedback subsystem, the efficiency of the optimized model is compared and analyzed. The results illustrate the improvement path of the dynamics model of the building system, and based on energy calculation, analyze the sustainable model parameter changes before and after, verify the effect of feedback optimization, and provide a reference direction for the deep coupling of the system dynamics model and the building system.

Keywords Building system; System dynamics; Full life cycle; Energy and sustainability

---

低碳可持续建筑生态位理论模型构建研究

Research on the Niche Theoretical Model Construction of Low-carbon Sustainable Buildings

吴伟东，曹稳，宋永伟，周玉佳
Wu Weidong, Cao Wen, Song Yongwei, Zhou Yujia

吴伟东，安徽科技学院建筑学院，教授，副院长。
曹稳，安徽科技学院建筑学院，讲师。
宋永伟，安徽科技学院建筑学院，讲师。
周玉佳，安徽科技学院建筑学院，讲师。

摘要 本研究从建筑与生态系统的相互影响关系进行考量，研究生态位理论在建筑领域的应用、建筑生态位理论与模型。对生态位维度、生态位宽度、生态位重叠、生态位态势、生态位适宜度进行初步研究；以案例为载体进行生态位评价方法初探，试图建立科学有效的优化模型，开展建筑领域的创新研究。

关键词 低碳可持续；建筑；生态位；评价方法

Abstract This study considers the interaction between architecture and ecosystem, studies the application of ecological niche theory in the architecture field, and the theory and model of architectural ecological niche. The dimensions of ecological niche, the width of ecological niche, the overlap of ecological niche, the situation of ecological niche, and the suitability of ecological niche were preliminarily studied. The case is taken as the carrier to explore the ecological niche evaluation method, trying to establish a scientific and effective optimization model and carry out innovative research in the field of architecture.

Keywords Low-carbon sustainable; Buildings; Ecological niche; Evaluation method

用于低碳建筑供暖的复合热源热泵系统敏感性分析

Sensitivity Analysis of a Solar-assisted Air Source Heat Pump for Space Heating System in Low-carbon Buildings

张少良，刘淑丽
Zhang Shaoliang, Liu Shuli

张少良，北京理工大学，博士研究生。
刘淑丽，北京理工大学，教授，博士生导师。

摘要 为降低建筑供暖系统碳排放，本文提出一种耦合混合蓄热的太阳能和空气复合热源热泵供暖系统；搭建了系统实验平台，测试并研究了系统在不同模式下的热力特性；基于实测数据，建立了系统节能性、环保性和经济性评价数学模型；选取中国北方六个典型城市作为研究对象，研究了太阳能保证率和太阳能热泵模式制热性能系数对系统节能性、环保性和经济性的影响。引入敏感性系数作为评价指标，探究了影响系统节能性、环保性和经济性的关键因素。

关键词 低碳建筑；供暖；复合热源热泵；敏感性分析

Abstract To reduce the carbon dioxide emissions of space heating systems in low-carbon buildings, this study proposes an indirect expansion solar-assisted air source heat pump (IX-SAASHP) with hybrid thermal energy storage. An experimental prototype is developed and tested to investigate the performance of the IX-SAASHP at different modes. Based on the experimental results, an efficient economic-environmental analysis mode of the IX-SAASHP system is established. This study selects six typical cities in China for techno-economic and environmental analysis. The influence of solar fraction and solar heat pump mode' s coefficient of performance on the techno-economic and environmental performance of the IX-SAASHP system is investigated. The key factors influencing the techno-economic and environmental performance of the IX-SAASHP system are explored using sensitivity indices.

Keywords Low-carbon building; Space heating; Solar-assisted air source heat pump; Sensitivity analysis

---

西藏地区商业建筑全生命周期碳排放测算及减碳策略研究

Study on Lifecycle Carbon Emission Calculation and Carbon Reduction Strategy of Commercial Buildings in Xizang

张海滨，袁首，李睿晨，何嘉琦，谢冲
Zhang Haibin, Yuan Shou, Li Ruichen, He Jiaqi, Xie Chong

张海滨，重庆大学建筑城规学院，副教授。
袁首，重庆大学建筑城规学院，硕士研究生。
李睿晨，何嘉琦，谢冲，重庆大学建筑城规学院，本科生。

摘要 西藏地区常规能源匮乏，建筑碳排放量较高，在“双碳”目标背景下，利用藏区丰富的太阳能、风能、地热能等可再生能源，进行低碳建筑设计具有必要性。本研究以藏区商业建筑为例，通过分析“台达杯”一等奖作品“藏街隐屋”的减碳技术体系，基于全生命周期评价法，对基地内商业建筑进行碳排放计算，并利用 DeST 软件进行建筑能耗模拟。结果表明，使用太阳能等技术并综合全生命周期减碳技术，可使建筑碳排放量大幅降低（约 65.73%）。

关键词 西藏地区；商业建筑；可再生能源；碳排放；全生命周期

Abstract Due to the lack of conventional energy and the high carbon emissions of buildings in Xizang, under the background of the dual-carbon target, it is necessary to use the rich solar, wind, geothermal and other renewable energy sources in Xizang to carry out low-carbon building design. This study takes commercial buildings in Xizang as an example. By analyzing the carbon reduction technology system of the first prize design of “Xizang Street Hidden House” in the International Solar Building Design Competitions, based on the lifecycle evaluation method, calculating the carbon emissions of commercial buildings in the base, and simulating the energy consumption of buildings with the DeST software. The results show that the use of solar energy and other technologies can greatly reduce carbon emissions during the operation stage of buildings. The comprehensive carbon reduction technology of the whole life cycle can reduce the carbon emissions of buildings by about 30%.

Keywords Xizang; Commercial buildings; Renewable energy; Carbon emissions; Lifecycle

CBEC：城市建筑能碳仿真模拟平台

CBEC: A Novel and Efficient Simulation Platform for Evaluating City-Scale Building Energy Use and Carbon Emissions

王晓宇，王超，田帅，耿跃云，石邢
Wang Xiaoyu, Wang Chao, Tian Shuai, Geng Yueyun, Shi Xing

王晓宇，田帅，同济大学建筑与城市规划学院，博士研究生。
王超，同济大学建筑与城市规划学院，博士后。
耿跃云，广联科技股份有限公司，部门总监。
石邢，同济大学建筑与城市规划学院，教授，博士生导师。

摘要 开发了一种新型高效的城市建筑能碳仿真模拟平台（City Building Energy and Carbon，CBEC），可实现城市尺度的建筑能耗和碳排放计算。CBEC 依据国内首部团体标准《城市建筑能耗及运行碳排放计算方法导则》开发。CBEC 平台可以为在城市规划设计中实现低碳节能的目标提供基础科学计算手段，为城市电力系统规划和区域能源站建设提供负荷需求数据，为城市能源政策制定者提供不同建筑节能措施的节能潜力分析报告和建议等。

关键词 城市建筑能耗；City Building Energy and Carbon (CBEC)；城市建筑碳排放；建筑节能；仿真模拟

Abstract City Building Energy and Carbon (CBEC), a novel and efficient simulation platform, focuses on evaluating and analyzing city-scale building energy use and carbon emissions is developed. CBEC is developed in accordance with the *Guidelines for Calculating Energy Consumption and Operational Carbon Emissions of Urban Buildings*, the first of its kind in China. CBEC provides a suite of features and application scenarios such as the low carbon urban planning and design for urban planners, urban electric system planning and district energy system planning for city energy managers, the energy saving potentials of different building energy efficiency initiatives for energy consultants and policy-makers, etc.

Keywords City-scale building energy use; City Building Energy and Carbon (CBEC); City-scale building carbon emissions; Building energy efficiency; Simulation

---

农村建筑群太阳能光伏潜力的时间特性分析

Analysis of the Temporal Characteristics of Solar Photovoltaic Potential in Rural Building Clusters

刘江，彭昌海
Liu Jiang, Peng Changhai

刘江，东南大学建筑学院，博士研究生。
彭昌海，东南大学建筑学院，教授，博士生导师。

摘要 为揭示农村建筑群太阳能光伏潜力的时间特性，本文以南京市的一个典型村庄为例，分析了农村建筑群屋顶和立面上年、月以及日内光伏潜力的变化特征。结果表明：农村建筑群的南、北向屋顶以及南、西向立面年光伏潜力最高，其占比超过建筑群年光伏总潜力的 70%。其中，南向立面的月光伏潜力变化趋势与另外三者相反，其在 5—7 月最小，而在 10—12 以及 3 月最大。西向立面的日内光伏潜力峰值时间出现最迟。通过合理调度不同建筑表面的光伏系统，可实现更稳定的农村建筑群全年光伏输出。研究结果为农村光伏政策的制定和系统部署提供了理论支撑。

关键词 农村；建筑群；光伏潜力；时间特性

Abstract To reveal the temporal characteristics of the solar photovoltaic (PV) potential of rural building clusters, this study analyzed the variations of annual, monthly and intra-day PV potential on the rural rooftops and facades by taking a typical village in Nanjing as a case study. The results showed that the annual PV potential of the south- and north-facing rooftops and the south- and west-facing facades are the highest, accounting for more than 70% of the total annual PV potential of the village. The monthly PV potential of the south-facing facade has the opposite trend to the other three, with the smallest from May to July and the largest from October to December and March. The peak intra-day PV potential of the west-facing facade occurs at the latest. A more stable year-round PV output of rural building clusters can be achieved by rationally dispatching PV systems that are installed on different building surfaces. The results provide theory support to rural PV policymaking and system deployment.

Keywords Rural areas; Building cluster; Photovoltaic potential; Temporal characteristics

基于 Ladybug + Dragonfly 性能模拟的科技园低能耗空间结构研究——以长三角地区为例

Research on Low-energy Space Structures of Science and Technology Parks based on Ladybug + Dragonfly Performance Simulation— A Case Study of the Yangtze River Delta

丁明琦，叶海，张涵琪
Ding Mingqi, Ye Hai, Zhang Hanqi

丁明琦，张涵琪，同济大学建筑与城市规划学院，硕士研究生。
叶海，同济大学建筑与城市规划学院，副教授。

摘要 在"双碳"目标大背景下，我国科技园低碳化发展的趋势明显，科技园在规划设计时不仅要考虑合理的功能组合和体量布局，还应当考虑充分利用可再生能源，降低整体的能耗水平。本文利用 Ladybug+Dragonfly 平台，对多个长三角地区的科技园实例进行太阳能利用潜力分析和区域建筑能耗模拟，探究圈层式布局和非圈层式布局下科技园整体能源效益的表现，发现非圈层式布局的科技园在设计实践中有更高的太阳能利用潜力，而在同类型布局中布局紧凑、建筑高度相对平均和建筑单体体型系数较小的科技园整体能耗水平更低。

关键词 科技园；空间结构；Ladybug+Dragonfly；太阳能利用潜力

Abstract Under the background of carbon peaking and carbon neutralization targets, the trend of low-carbon development of Science and Technology Park(STP) in China is obvious. When planning and designing STP, we should not only consider the reasonable function combination and volume layout, but also consider the full use of renewable energy and reduce the overall energy consumption level. In this paper, Ladybug+Dragonfly platform is used to analyze the solar energy utilization potential and the regional building energy consumption of several examples of STP in the Yangtze River Delta region, exploring the performance of the overall energy efficiency of STP under the circular layout and non-circular layout. It is found that the non-circular layout of STP has higher solar energy utilization potential in design practice, and in the same type of layout, the one with more compact layout, relatively average building height and small building shape coefficient has lower overall energy consumption.

Keywords Science and technology park; Space structure; Ladybug+Dragonfly; Utilization potential of solar energy

---

武汉城市圈不透水面扩张与地表温度多时空耦合研究

Study on the Spatio-Temporal Coupling of Impervious Surface Expansion and Surface Temperature in the Wuhan Urban Agglomeration

吴标平，周雪帆
Wu Biaoping, Zhou Xuefan

吴标平，华中科技大学建筑与城市规划学院，硕士研究生。
周雪帆，华中科技大学建筑与城市规划学院，硕士生导师，副教授。

摘要 城市化进程导致城市不透水面积（IS）不断增加，进而加剧城市热岛效应。研究 IS 与城市地表温度（LST）的时空变化关系，对指导城市圈城市化进程中 IS 的扩展具有重要意义。本文以武汉城市圈为例，采用多时空轨迹和标准差椭圆考察 LST 和不透水表面分数（ISF）时空动态。研究结果表明，武汉城市圈的 IS 扩张具有明显的方向性，总体朝东，LST 扩张方向性不明显，总体朝西，两者移动总体都朝向指向武汉市中心。LST 和 ISF 的加权中心距离呈减少趋势，移动夹角缩小，总体耦合态势不断加强。随年份增长，标准差椭圆的方位角、方向性和离散性都逐渐增大，LST 对 ISF 变化的敏感度逐渐趋于一致。

关键词 地表温度；不透水表面；轨迹分析；武汉城市圈

Abstract Urbanization processes lead to a continuous increase in impervious surface area (IS), exacerbating the urban heat island effect. Studying the spatio-temporal relationship between IS and land surface temperature (LST) is crucial in guiding IS expansion during urbanization in urban clusters. Using multi-temporal trajectories and standard deviation ellipses, this study examines the spatio-temporal dynamics of LST and impervious surface fraction (ISF) in the Wuhan urban agglomeration. Findings reveal that IS expansion in Wuhan shows clear eastward directionality, while LST expansion has less apparent westward directionality, with both converging towards the center of Wuhan city. Weighted centers of LST and ISF exhibit decreasing distances and narrowing angular displacement, indicating increasing overall coupling. Over time, the azimuth, directionality, and variability of standard deviation ellipses gradually increase, and the sensitivity of LST to ISF changes tends to converge.

Keywords LST; Impervious surface; Trajectories analysis; Wuhan urban agglomeration

上海市青浦区乡村民宅低碳改造设计研究

Research on Low-carbon Renovation Design of Country Houses in Qingpu, Shanghai

常嘉诚，丁明琦，张涵琪
Chang Jiacheng,
Ding Mingqi,
Zhang Hanqi

摘要 在我国广大乡村地区存在着大量民宅，这些民宅由于地理、文化等方面的不同而各具特色。这些民宅虽然积累了当地多年来的建造经验，但往往没有节能技术的支撑。在“双碳”目标的大背景下，乡村民宅需要引入新的技术来帮助其实现低碳节能化改造。本文以上海市青浦区的一种典型乡村民宅为例，将多种节能技术与建筑自身条件及建筑周边环境相结合，探讨节能技术在民宅改造中的使用，并用软件模拟辅助设计。

关键词 民居改造；低碳改造设计；夏热冬冷；太阳能

Abstract There are a large number of private houses in the vast rural areas of our country. These houses have their own characteristics due to different geographical and cultural differences. Although these houses have accumulated many years of local construction experience, they are usually not supported by energy-saving technology. In the context of the carbon peaking and carbon neutrality goals, rural houses need to introduce new technologies to help them achieve low-carbon and energy-saving transformation. This paper takes a typical rural house in Qingpu District, Shanghai as an example to discuss the use of energy-saving technology in the renovation of residential buildings by combining a variety of energy-saving technologies with the building's own conditions and the surrounding environment and using software simulation to assist the design.

Keywords Renovation design of country houses; Low-carbon renovation design; Hot summer and cold winter zone; Solar energy

常嘉诚，同济大学，硕士研究生。
丁明琦，同济大学，硕士研究生。
张涵琪，同济大学，硕士研究生。

广东省某高校教学和宿舍楼的用能及光伏发电匹配研究

Study on the Matching of Energy Consumption and Photovoltaic Power Generation System in a University Teaching and Dormitory Building in Guangdong Province

张琬琨，唐海达，李春莹 *
Zhang Wankun,
Tang Haida,
Li Chunying

摘要 高校建筑用能密度高，校园节能减排是“双碳”目标的重要内容。本文以广东省某高校教学和宿舍楼为例，研究建筑光伏一体化对于减少校园建筑能耗的作用。利用 Ladybug + Honeybee 建立教学、宿舍楼的用能模型和屋面光伏系统模型，并分别模拟全年能耗和年发电量。对光伏系统年发电量与建筑能耗的逐时匹配情况进行分析，结果表明屋顶光伏可弥补两类建筑群 29.96% 的年能耗，高校节能需要多种手段进行系统改造。

关键词 校园建筑；建筑节能；光伏建筑；能耗模拟

Abstract The energy consumption density of university is generally high. The related energy saving and emission reduction is an important part of the carbon peaking and carbon neutrality goals. This paper takes university teaching, dormitory buildings in Guangdong Province as an example to study the effect of building photovoltaic integration in reducing building energy consumption and reducing carbon emissions on campus. Ladybug + Honeybee is used to model the energy usage of teaching building, dormitory building and roof photovoltaic system, simulate the annual energy consumption and power generation respectively. The time-by-time matching of annual building energy consumption and power generation of PV systems were analyzed, and the results illustrated that rooftop photovoltaics can offset 29.96% of the annual energy consumption in both types of building complexes, the energy saving of the university needs multiple means for system modification.

Keywords Campus building; Building energy efficiency; Photovoltaic buildings; Energy consumption simulation

张琬琨，深圳大学建筑与城市规划学院，硕士研究生。
唐海达，深圳大学建筑与城市规划学院，深圳市医养建筑重点实验室（筹建启动），研究员，硕士生导师。
李春莹，深圳大学建筑与城市规划学院，深圳市医养建筑重点实验室（筹建启动），副教授。

装配式竹建筑减碳特性模拟研究

Simulation Study on Carbon Reduction Characteristics of Prefabricated Bamboo Buildings

范旭悦，刘紫琪，刘大龙，韩雨菲，王国乾
Fan Xuyue, Liu Ziqi, Liu Dalong, Han Yufei, Wang Guoqian

摘要 随着绿色革命的发展，竹材因其固碳性能良好、造价低廉和能源效率高等优点吸引了全球的关注。基于传统竹建筑连接方式发展的钢制套筒预制装配节点，可以在满足批量预制及装配化的同时实现后续的再组装和重复使用。基于此项新技术专利进行建筑设计，利用斯维尔和 PKPM 建筑性能模拟软件对建筑的建筑碳排放及热性能进行模拟分析。研究表明，该住宅物化阶段碳排放总量 735.66 t，使用阶段碳排放总量 1421.24 t，建筑全生命周期单位面积碳排放量 59 kg/(m²·a)，对比钢筋混凝土参照建筑减碳效果接近 50%。

关键词 竹建筑；建筑低碳设计；装配技术；低碳建筑；建筑节能

Abstract With the development of the green Revolution, bamboo has attracted global attention because of its good carbon sequestration performance, low cost and high energy efficiency. The steel sleeve prefabricated assembly node developed based on the traditional bamboo building connection can meet the requirements of batch prefabrication and assembly, and realize the subsequent reassembly and reuse. Based on the new technology patent, the architectural design is carried out, and the building carbon emission and thermal performance of the building are simulated and analyzed by Swire and PKPM building performance simulation software. The total carbon emission of the house in the physical and chemical stage is 735.66 t, the total carbon emission in the use stage is 1421.24 t, and the carbon emission per unit area of the building in the whole life cycle is 59 kg/(m².a), which is close to 50% compared with the carbon reduction effect of the reinforced concrete reference building.

Keywords Bamboo architecture; Low-carbon building design; Assembly technology; Low-carbon buildings; Building energy conservation

范旭悦，西安建筑科技大学建筑学院，硕士研究生。
刘紫琪，西安建筑科技大学信息与控制工程学院，硕士研究生。
刘大龙，西安建筑科技大学建筑学院，副教授。
韩雨菲，西安建筑科技大学建筑学院，硕士研究生。
王国乾，西安建筑科技大学建筑学院，工程师，博士研究生。

兰州市青城镇三生空间低碳优化与净碳排量分析

Low-carbon Optimization and Net Carbon Emission Analysis of the Production, Living and Ecological Space in Qingcheng Town, Lanzhou City

张月，王天鹏
Zhang Yue, Wang Tianpeng

摘要 为解决乡村旅游大规模开发和快速建设引发的碳排放问题，碳排的定量化研究逐渐成为研究热点。本文从低碳建设、政府干预、公众参与几个层面对兰州市国家级历史文化名镇——青城镇进行低碳优化分析，并基于实地走访及 2022 年《甘肃发展年鉴》相关数据，利用排放因子法、交通行驶里程计算法、固碳系数法等计算对比镇内生产空间、生活空间、生态空间低碳优化前后碳排量和碳汇量。结果表明，优化后镇内碳排量减少、碳汇量增加，净碳排放量降低率达 30.06 %。

关键词 三生空间；碳排碳汇；低碳建设；公众参与；青城镇

Abstract In order to solve the problem of carbon emissions caused by large-scale development and rapid construction of rural tourism, the quantitative research of carbon emissions has gradually become a research hotspot. This paper analyzes the low-carbon optimization of Qingcheng Town, a national historical and cultural town in Lanzhou, from the aspects of low-carbon construction, government intervention and public participation. Based on the field visit and the relevant data of Gansu Development Yearbook, the paper uses the emission factor method, traffic mileage calculation method, carbon sequestration coefficient method and other methods to calculate and compare the carbon emissions and carbon sinks before and after the low-carbon optimization of the production space, living space and ecological space in the town. The results show that after optimization, the carbon emission in the town is reduced, the carbon sink is increased, and the net carbon emission is reduced by 30.06%.

Keywords Production, living and ecological space; Carbon emission and sink; Low-carbon construction; Public participation; Qingcheng Town

张月，兰州交通大学建筑与城市规划学院，硕士研究生。
王天鹏，兰州交通大学建筑与城市规划学院，教授，硕士生导师。

热力学视角下的建筑气候适应性评估与设计方法

Assessment and Design Method of Building Climate Adaptability from the Perspective of Thermodynamics

何美婷，李麟学 *，张琪，张广益
He Meiting, Li Linxue, Zhang Qi, Zhang Guangyi

何美婷，同济大学建筑与城市规划学院，博士后。
李麟学，同济大学建筑与城市规划学院，教授，博士生导师。
张琪，同济大学建筑与城市规划学院，博士研究生。
张广益，同济大学建筑与城市规划学院，博士研究生。

摘要 在被动式建筑的设计中，除了要遵循传统乡土建筑气候适应性的经验方法以外，还需要对当代建筑的生物气候设计措施进行实质性的验证，有效地开发出一种由经验和热力学分析方法组成的结合式方法。因此，本文将以浙江一文学馆设计为例，从能量和热力学的观点来重新审视建筑气候适应性，提出采用平衡点温度的计算来分析和衡量建筑采用被动式设计方法的可能性，从而评估在特定的时间段内，建筑物是否能够达到环境的热平衡。由此，可以更清楚地了解设计初期所设计的建筑形体的能量需求，并据此进一步对后续的设计进行气候适应性的优化。

关键词 气候适应性；热力学生态建筑；被动式建筑；平衡点温度

Abstract This paper will take the design of a literature museum in Zhejiang Province as an example, re-examine the climate adaptability of buildings from the perspective of energy and thermodynamics, and further explain the climate adaptability analysis process of local buildings. Among them, it is proposed to use the calculation of equilibrium point temperature to analyze and measure the possibility of building adopting passive design method, so as to evaluate whether the building can reach the thermal balance of the environment in a specific period of time.

Keywords Climate adaptability; Thermodynamic ecological architecture; Passive building; Balance point temperature

---

随需而变：空间互补视角下的北京居住院落更新策略

Changing with Demands: Exploration on the Renewal of Residential Courtyards in Beijing from the Perspective of Spatial Complementary

王璐璐，姚振，李佳乐，梁佳，董宏
Wang Lulu, Yao Zhen, Li Jiale, Liang Jia, Dong Hong

王璐璐，姚振，李佳乐，北京建筑大学建筑与城市规划学院，硕士研究生。
梁佳，北京建筑大学建筑与城市规划学院，讲师。
董宏，绿色建筑与节能技术北京市重点实验室，北京建筑大学建筑与城市规划学院，教授。

摘要 大杂院是北京老城院落的普遍现状，由于人口密度大、基本生活设施不足，居住环境亟须改善。本文基于人群 - 时间 - 行为模式的分析，提出基于虚空间互补转换的更新策略。首先，通过剖析各需求的空间构成及使用性质，对其“虚空间”进行提取；其次，依据各人群在不同时段的需求变化，对可互补的“虚空间”进行组合，以满足有限空间内多种功能的转变，从小尺度着手满足现在、未来人群的基本生活需求，以期实现北京老城院落居住空间的生活宜居。

关键词 北京老城；院落更新；基本生活需求；可变空间；互补

Abstract The current miscellaneous courtyard has become the general status of the Beijing Old City Courtyard. Due to the large population density and insufficient basic living facilities, the living environment needs to be improved urgently. The article is based on the analysis of the crowd-time-behavioral mode, proposes an update strategy based on virtual space complementary transformation. Firstly, by analyzing the composition and use of the space of each demand space, the "void space" is extracted; Secondly, according to the changes in the needs of each population in different periods, the complementary "void" is combined to meet the changes in a variety of functions in the limited space. Starting from a small scale, we aim to meet the basic living needs of the present and future population, in order to achieve livable living in the courtyard of the old city of Beijing.

Keywords Beijing old city; Courtyard renewal; Life needs; Variable space; Complementary

西安市海绵型街谷绿地建筑空间格局碳污影响研究

Study on the Carbon Pollution Impact of Sponge Street Valley Green Space and Building Spatial Pattern in Xi'an City

安琦，马西娜，王梦媱，李金铭
An Qi，Ma Xina，Wang Mengyao，Li Jinming

安 琦，长安大学，硕士研究生。
马西娜，长安大学，副教授，硕士研究生导师。
王梦媱，长安大学，硕士研究生。
李金铭，长安大学，硕士研究生。

摘要 为改善城市生态环境，本文以西安市海绵型街谷为研究对象，选取微气候因子与绿地-建筑空间格局指标，采用实地监测的方法，研究在夏季气象条件下不同指标对街道内的碳污扩散影响规律，并以此提出具有可操作性的海绵型街道减污降碳景观格局规划措施，为海绵城市建设提供研究支持。

关键词 绿地景观格局；海绵型街谷；$PM_{2.5}$；CO_2

Abstract In order to improve the urban ecological environment, the study take Xi'an Sponge Street Valley as the research object, and selects microclimatic factors and green building space pattern indicators, and uses field monitoring methods to investigate the impact of different indicators on the diffusion of carbon pollution in streets under summer meteorological conditions. The study aims to propose an operational sponge street pollution reduction measure and landscape pattern carbon reduction planning measure, and to provide research support for the construction of sponge cities.

Keywords Green space landscape pattern; Spongy street; $PM_{2.5}$; CO_2

高密度城市住区建筑光伏适宜空间分布特征
——以武汉为例

Suitable Spatial Distribution Characteristics of PV in Residential Buildings in High-density Urban Blocks—A Case Study of Wuhan

徐燊，桑梦城，杨翰，侯俊豪
Xu Shen, Sang Mengcheng, Yang Han, Hou Junhao

徐燊，华中科技大学，教授。
桑梦城，杨翰，侯俊豪，华中科技大学，硕士研究生。

摘要 顺应“双碳”目标与“十四五”规划，从应用角度考虑城市住区分布式光伏利用潜力评估对于城市节能减碳具有重要意义。本文在现有利用潜力评估的研究基础上，将空间遮挡因素较少、辐射量较为丰富的空间定义为光伏适宜空间，并对其开展评估，评估包括武汉市典型街区建筑外立面太阳辐射接收量模拟、光伏适宜空间分布特征可视化、城市住宅街区规模化光伏应用技术参考与设计引导，助力实现城市低碳发展。

关键词 “双碳”目标；建筑分布式光伏应用；光伏适宜空间；形态因素

Abstract In accordance with the new requirements of the carbon peaking and carbon neutrality goals and the 14th Five-Year Plan, it is of great significance to consider the potential assessment of distributed photovoltaic utilization in urban residential areas from the perspective of application for energy saving and carbon reduction. Based on the existing research on the evaluation of the utilization potential, this paper defines the space with fewer spatial shielding factors and more abundant radiation as the suitable photovoltaic space for evaluation. The evaluation includes the simulation of solar radiation reception on the facade of typical blocks in Wuhan City, visualisation of suitable photovoltaic spatial distribution characteristics, technical reference and design guidance for large-scale photovoltaic application in urban residential blocks, contributing to low-carbon urban development.

Keywords Carbon Peaking and Carbon Neutrality Goals; Distributed photovoltaic applications in buildings; Photovoltaic suitable space; Morphological factor

BIM技术在施工管理中的应用

The Application of BIM Technology in Construction Management

陈婷，乌云巴根，佘洁卿
Chen Ting, Wuyun Bagen, She Jieqing

摘要 传统施工管理方法存在数据协同率低、施工现场布置不规范、资源配置不合理、施工进度计划不科学、可视化效果差等问题。在数字化时代背景下，运用智慧建造施工管理方法，可以实现规范化管控和精细化管理施工过程。本文以龙岩市中国移动上杭生产调度楼工程为例，采用BIM技术的智慧建造施工管理方法进行模拟施工，利用BIM模型进行优化图纸、资源用量配置、编制施工进度计划、施工场地布置等方面的应用，科学合理控制施工进度、质量、成本、安全、环境，实现工程项目智慧化、精准化、可视化，为建筑工程的施工管理工作提供参考依据。

关键词 BIM技术；智慧建造；建筑施工管理；可视化

Abstract Traditional construction management methods have problems e.g. low data coordination rate, irregular construction site layout, unreasonable resource allocation, unscientific construction schedule, and lack of visual expression. In the digital era, intelligent construction management methods can standardize the control and refine the management of the construction process. This article takes Longyan City China Mobile Shanghang production dispatch building project as an example, adopts the intelligent construction management method of BIM technology to simulate the construction，and uses BIM model to optimize the application of drawings, resource consumption allocation, construction schedule plan, construction site layout and etc. It scientifically and reasonably controls the construction progress, quality, cost, safety and environment, it realizes the wisdom, precision and visualization of engineering projects, and provides a reference basis for the construction management.

Keywords BIM; Smart construction; Building construction management; Visualization

陈婷，华侨大学建筑学院，硕士研究生。
乌云巴根，华侨大学建筑学院，讲师。
佘洁卿，莆田学院土木工程学院，讲师。

景观评价方法对比及其在传统园林研究中的适用分析

Comparison of Landscape Evaluation Methods and Their Application in Traditional Garden Research

曹书璇，乌云巴根
Cao Shuxuan, Wuyun Bagen

摘要 随着中国特色的科学景观评价体系的构建，需要选取适宜的评价方法，体现中国特有的美学元素与环境观，构建具有中国特色和可操作性的评价方法。文章研读近年来国内外景观评价方法的相关文献，总结了语义差异法（SD法）、美景度评判法(SBE法)、层次分析法(AHP法)、比较评判法（LCJ法）、审美评价测量法（BIB-LCJ法）、人体生理心理指标测试法（PPI法）等主要评价方法的优缺点和适用案例，及其在中国传统园林的景观评价中的适用性探讨，为其他中国特色景观资源尤其是小尺度的景观评价提供参考。

关键词 中国传统园林；景观评价；心理物理学派；植物景观；对比分析

Abstract The construction of the scientific landscape evaluation system with Chinese characteristics requires the selection of appropriate evaluation methods to reflect China's unique aesthetic elements and environmental views, so that the methods have Chinese characteristics and operability. The article reviews the relevant literature on landscape evaluation methods at home and abroad in recent years, summarizes the advantages, disadvantages, and applicable cases of the main evaluation methods including SD method, SBE method, AHP method, LCJ method, BIB-LCJ method, and PPI method, and explores their applicability to the evaluation of traditional Chinese gardens. This provides reference for other Chinese characteristic landscape resources, especially small-scale landscape evaluation.

Keywords Traditional Chinese gardens; Landscape evaluation; Psychophysics; Plant landscape; Comparative analysis

曹书璇，华侨大学建筑学院，硕士研究生。
乌云巴根，华侨大学建筑学院，讲师。

“双碳”目标导向下的现代木构应用

The Application of Modern Wood Structure Under the Guidance of Carbon Peaking and Carbon Neutrality Goals

王石满，何文晶
Wang Shiman,
He Wenjing

王石满，山东建筑大学建筑城规学院，硕士研究生。
何文晶，山东建筑大学建筑城规学院，教授。

摘要 2020年我国提出“双碳”目标后，关于建筑运行碳排放等一系列政策文件和规范标准相继出台。在建筑全生命周期碳排放中，选择绿色低碳建材具有重要意义，同时，木材的物理性能对于减少建筑能耗方面也能发挥重要作用。研究现代木构的发展现状和现代木构的应用对于减碳环保具有重要意义。本文简要介绍我国现代木结构的现状，选取国外案例，对现代木构在不同建筑类型的应用以及木材在建筑中的优势进行分析，以期为我国木结构建筑创作提供相关参考。

关键词 现代木结构；零碳建筑；零能耗建筑；“双碳”

Abstract Following China's goals of carbon peaking and carbon neutrality by 2020, a series of policy documents and normative standards on carbon emissions from building operations have been issued. The choice of green and low-carbon building materials is of great significance in the carbon emissions of the whole life cycle of buildings, and the physical properties of wood can also play an important role in reducing the energy consumption. It is of great significance to study the development status of modern wood structure and the application of modern wood structure for carbon reduction and environmental protection. This paper briefly introduces the current situation of modern wood structure in China, selects foreign cases, and analyzes the application of modern wood structure in different building types and the advantages of wood in construction, in order to provide relevant references for the creation of wood structure in China.

Keywords Modern wood construction; Zero-carbon building; Zero-energy consumption building; Carbon peaking and carbon neutrality

---

农机电池介入下农村居住建筑柔性负荷优化调控

Optimal Control of the Flexible Load of Rural Residential Buildings with the Intervention of Agricultural Machinery Battery

罗西，石文策，杨燕子
Luo Xi, Shi Wence,
Yang Yanzi

罗西，西安建筑科技大学，副教授。
石文策，西安建筑科技大学，硕士研究生。
杨燕子，西安建筑科技大学，硕士研究生。

摘要 随着农业电气化的快速发展，蓄电型农机在农村愈发普及，利用闲置期的农机储能电池协助居住建筑用能调蓄将成为解决供需动态匹配问题的新途径。本研究以陕西关中地区典型农村为例开展调研，以考虑农机电池介入为前提条件，建立求解以经济性为目标的农村居住建筑柔性负荷优化调控模型。结果表明：①农机电池资源和柔性负荷在时间分布上呈现出明显的季节性变化特征；②农机电池的有效蓄放使典型农户的光伏消纳率在夏季农忙、夏季农闲、过渡季农闲、冬季农闲典型日分别提升0.15%、5.99%、3.60%和8.41%。

关键词 需求侧响应；柔性负荷；农业机械；户用光伏

Abstract With the rapid development of agricultural electrification, the electric storage agricultural machinery is becoming increasingly in rural areas. Using the energy storage battery of agricultural machinery in the idle period to assist the energy storage of residential buildings will become a new way to solve the problem of dynamic matching between supply and demand. This study takes typical rural areas in the Guanzhong of Shaanxi region as an example to conduct research, taking the intervention of agricultural machinery and batteries as the prerequisite, and establishes an optimization and control model for flexible load of rural residential buildings with economic objectives. The results showed that: ① The time distribution of agricultural machinery battery resources and flexible loads shows obvious seasonal variation characteristics; ② The effective storage and discharge of agricultural machinery batteries can increase the photovoltaic consumption rate of typical farmers by 0.15%, 5.99%, 3.60%, and 8.41% on typical days of summer farming, summer fallowing, transition season fallowing, and winter fallowing, respectively.

Keywords Demand response; Flexible load; Agricultural machinery; Household photovoltaic system

低碳导向下的夏热冬冷地区住宅组团形式多目标优化设计

Multi-objective Optimization Design of Residential Group Form in Hot Summer and Cold Winter Areas Under the Guidance of Low Carbon

徐燊，杨翰，侯俊豪，李鉴儒
Xu Shen，Yang Han，Hou Junhao，Li Jianru

徐燊，华中科技大学建筑与城市规划学院，教授。
杨翰，侯俊豪，李鉴儒，华中科技大学建筑与城市规划学院，硕士研究生。

摘要 随着中国城市化的高速发展，住宅小区大量出现并成为能源消耗的一大因素。本文以符合城市发展理念的政策文件为基础，对组团尺度的住宅建筑，进行能耗与光伏潜力的多目标优化设计。通过犀牛 GH 平台生成住宅的组团原型，并对其结果进行筛选，获得满足城市设计规范要求与经济性的住宅原型结果。综合遗传算法的帕累托解集得出住宅组团尺度下满足能耗较低且光伏潜力较大的组团形式、南立面窗墙比、建筑长宽比与建筑平均高度等相关参数，为城市化的发展提供相应指引。

关键词 低碳城市；多目标优化；建筑能耗；光伏潜力；住宅组团

Abstract With the rapid development of urbanization in China, a large number of residential quarters have appeared and become a major factor of energy consumption. Based on the policy documents in line with the concept of urban development, this paper carries out multi-objective optimization design of energy consumption and photovoltaic potential for residential buildings with group scale. Through rhinoceros GH platform, the residential group prototype is generated, and the results are screened, and the residential prototype results that meet the requirements of urban norms and design economy are obtained. The Pareto solution set of comprehensive genetic algorithm is used to obtain the relevant parameters such as the group form, the window-wall ratio of the south facade, the length-width ratio of the building and the average height of the building, which meet the requirements of low energy consumption and great photovoltaic potential under the scale of residential group, and provide corresponding guidance for the development of urbanization.

Keywords Low-carbon cities; Multi-objective optimization; Building energy consumption; Photovoltaic potential; Residential group

---

基于 CiteSpace 的垂直绿化研究进展及热点可视化分析

Research Progress on Vertical Greening Based on CiteSpace and Visual Analysis of Hot Spots

汪雨，冉茂宇
Wang Yu，Ran Maoyu

汪雨，华侨大学建筑学院，硕士研究生。
冉茂宇，华侨大学建筑学院，教授。

摘要 面对“双碳”目标，城市建筑碳排放问题急需解决。在有限的城市空间中，垂直绿化作为一种建筑表皮自然缓解策略应运而生。为了解当前垂直绿化相关研究进展、热点等，基于中国知网（CNKI）数据库，运用文献计量法和 CiteSpace 可视化软件分析法，分析了近 20 年来垂直绿化相关研究中英文文献的发文量、研究机构及作者、热点及演进脉络时间线等内容，以期能为相关研究提供基础参考。结果表明，垂直绿化研究机构主要为高校，热点集中在热性能、节能、生态等方面，研究趋势近年来趋于稳定。

关键词 垂直绿化；CiteSpace；研究进展；可视化分析

Abstract In view of the carbon peaking and carbon neutrality goals, the carbon emission problem of urban buildings needs to be solved urgently. In the limited urban space, vertical greening has emerged as a natural mitigation strategy for the building skin. In order to understand the current research progress and hot spots related to vertical greening, based on CNKI database, using bibliometric method and CiteSpace visualization software analysis method, the publication volume, research institutions and authors, hot spots and evolution timeline of vertical greening related research in the past 20 years were analyzed, in order to provide basic reference for related research. The results show the vertical greening research institutions are mainly universities, and the hot spots are concentrated in thermal performance, energy saving, ecology, etc., and the research trend has stabilized in recent years.

Keywords Vertical greening；CiteSpace；Research progress；Visual analytics

景观对校园室外热环境的影响研究——以成都市沙河源小学为例

Research on the Impact of Landscape on Campus Outdoor Thermal Environment—Taking Shaheyuan Primary School in Chengdu as an Example

杨昕宜，欧阳金龙，高庆龙
Yang Xinyi,
Ouyang Jinlong,
Gao Qinglong

杨昕宜，四川大学，硕士研究生。
欧阳金龙，四川大学，教授。
高庆龙，四川大学，教授。

摘要 校园是儿童长时间活动的场所，其室外热环境对儿童身心健康的影响不容忽视。景观作为室外环境的重要组成部分，对热环境有一定影响。本研究以成都市沙河源小学为例，采用热环境现场测试、问卷访谈及观察记录的方法，分析热环境参数、评价儿童热舒适。结果表明，不同景观空间类型对室外热环境的影响不同；利用建筑灰空间打造具吸湿性材料的构筑物与绿植组合的景观等方法可显著改善夏热冬冷地区校园室外热环境质量；热舒适受太阳辐射影响大，需基于太阳辐射规律优化设计校园景观。

关键词 室外热环境；热舒适；景观；小学

Abstract Campus is a place for children' s long-time activities, and the influence of its outdoor thermal environment on children' s physical and mental health should not be ignored. As an important part of the outdoor environment, landscape has a certain influence on the thermal environment. Taking Shaheyuan Primary School in Chengdu as an example, thermal environment field tests, questionnaire interviews and observation records were used to analyze thermal environment parameters and evaluate children' s thermal comfort. The results show that different landscape space types have different impacts on the outdoor thermal environment; using the grey space of the building to create a landscape with hygroscopic materials and greenery can significantly improve the quality of the outdoor thermal environment of campuses in hot summer and cold winter areas; thermal comfort is greatly affected by solar radiation, and campus landscapes need to be optimized based on the laws of solar radiation.

Keywords Outdoor thermal environment; Thermal comfort; Landscape; Primary school

---

基于自然通风的高层公寓厨房空间形态优化研究

Study on Morphology Optimization of Kitchens in High-Rise Apartment Based on Natural Ventilation

徐燊，李鉴儒，王明豪，杨翰
Xu Shen, Li Jianru,
Wang Minghao, Yang Han

徐燊，华中科技大学建筑与城市规划学院，教授。
李鉴儒，华中科技大学建筑与城市规划学院，硕士研究生。
王明豪，华中科技大学建筑与城市规划学院，博士研究生。
杨翰，华中科技大学建筑与城市规划学院，硕士研究生。

摘要 厨房油烟作为住宅室内的主要污染物，包含大量细颗粒物与有毒有害气体物质，严重威胁人体健康。高层公寓厨房具有空间小、标准化的特点。自然通风是改善厨房环境的有效手段。本文通过对武汉典型公寓平面的调研总结，通过 CFD 模拟，研究公寓空间形态对厨房自然通风的影响规律，以厨房开敞度和位置、公寓面宽和进深为变量，以厨房及起居区域的平均风速为评价指标，提出公寓厨房空间形态优化策略。

关键词 高层公寓；厨房；自然通风；CFD 模拟；形态优化

Abstract As the main pollutants in residential buildings, cooking pollutants contain a large amount of fine particulate matter and toxic and harmful gas substances, which seriously threaten people' s health. High-rise residential kitchens are characterized by small space and standardization. Natural ventilation is an effective way to improve the kitchen environment. By summarizing the research on typical apartment levels in Wuhan area, influence of morphology optimization of apartment on the natural ventilation environment of kitchen was studied by CFD simulation. The kitchen openness and location, apartment width and depth were taken as variables, and the average wind velocity of kitchen and living area were taken as evaluation index in order to propose kitchen morphology optimization strategies.

Keywords High-rise apartments; Kitchen; Natural ventilation; CFD simulations; Morphology optimization

绿色低碳导向下基于农旅一体化的装配式农宅设计研究

Research on the Design of Prefabricated Rural Residence Based on the Integration of Agriculture and Tourism under the Guidance of Green and Low-Carbon

梅兰，阎阳，陈晓丹，梅森
Mei Lan, Yan Yang, Chen Xiaodan, Mei Sen

梅兰，河北建筑工程学院，讲师。
阎阳，河北建筑工程学院，副教授。
陈晓丹，河北建筑工程学院，讲师。
梅森，河北建筑工程学院，本科生。

摘要 本文基于张家口农村地区的调查研究，以装配式农宅为载体，探索适应当地农旅一体化发展特点的低能耗农宅建造与推广方案。选取位于沽源天鹅湖畔的田家营子村农宅为研究对象，提出从功能整合、户型设计、装配式建造到绿色建筑技术应用等方面的一体化集成设计策略，以期为该地区乡村绿色建筑设计提供一定参考依据。

关键词 绿色低碳；农旅一体化；装配式农宅；低能耗；设计策略

Abstract This paper investigates and studies the prefabricated rural residence of Zhangjiakou, explores the construction and promotion scheme of low energy consumption rural residence that adapts to the characteristics of local integrated development of agriculture and tourism. The author selects the rural residence located around the Swan Lake Tianjiayingzi Village of Guyuan county as the research object, puts forward the integrated design strategy of functional integration, house type design, prefabricated construction and green building technology application, in order to provide some reference for the design of rural green buildings in this region.

Keywords Green and low-carbon; Integration of agriculture and tourism; Prefabricated rural residence; Low energy consumption; Design strategy

---

商住混合街区形态与碳排放的关联性研究

Research on the Correlation Between Mixed Commercial And Residential Block Forms and Carbon Emissions

巨路伟，陈宏
Ju Luwei, Chen Hong

巨路伟，华中科技大学建筑与城市规划学院，硕士研究生。
陈宏，华中科技大学建筑与城市规划学院，教授，博士生导师。

摘要 在“双碳”目标下，碳排放量比重较大的建筑行业应当做出积极应对。目前，城市街区形态与碳排放的关联性研究以单一功能街区为主，对混合功能街区的研究较少。本文以武汉商住混合街区为研究对象，提取街区中与碳排放相关的形态因子，并通过 EnergyPlus 能耗模拟、线性回归分析研究街区形态对建筑运行阶段碳排放的影响。最后，研究得到街区商住比、容积率、建筑密度、体形系数与建筑碳排放的关系，为商住混合街区的低碳设计提供有效参考。

关键词 商住混合街区；空间形态因子；EnergyPlus；碳排放量

Abstract Under the background of the carbon peaking and carbon neutrality goals, the architectural industry, which has a large proportion of carbon emissions, should actively respond. At present, researches on the correlation between urban block forms and carbon emissions focus on single-function blocks. There is little research on mixed-function blocks. This paper takes the mixed commercial-residential blocks in Wuhan as the research object. EnergyPlus energy consumption simulation and linear regression analysis were used to study the effect of block form on carbon emissions during the building operation stage. Finally, the relationship between block commercial to residential floor area ratio, floor area ratio, building density, form factor and block's carbon emissions was quantified, which can provide good references for the low-carbon design of mixed-function blocks.

Keywords Mixed-use commercial and residential blocks; Spatial morphological factor; EnergyPlus; Carbon emissions

广州地区建筑屋顶光伏大面积淋水降温提效潜力研究

Research on the Efficiency Potential of Large-area PV Drenching and Cooling Methods on Building Rooftops in Guangzhou

汪博然，孟庆林，赵鹤阳
Wang Boran,
Meng Qinglin,
Zhao Heyang

汪博然，华南理工大学，硕士研究生。
孟庆林，华南理工大学，教授，亚热带建筑科学国家重点实验室副主任，中国建筑学会建筑物理分会理事长。
赵鹤阳，华南理工大学，硕士研究生。

摘要 光伏发电过程中积蓄的热量会导致光伏模块转换效率的降低，且严重影响光伏组件的寿命。为对比广州地区不同淋水方式对建筑屋顶光伏的降温提效潜力，本研究搭建独立光伏组件降温效果测试实验台，选取晴朗天气中太阳辐射超过 800 W/m^2 的时间段进行独立光伏发电系统的对比实验。结果表明，几种淋水方式降温效果明显，采用摇臂喷头淋水具有更大的提效潜力，持续淋水时光电转换效率增长率达 9.26%。进一步考虑屋顶光伏阵列的尺寸与喷头淋水射程等，摇臂喷头和旋转射线喷头更适宜实际安装使用。

关键词 建筑屋顶光伏；光伏组件提效；淋水降温；降温提效潜力

Abstract The heat accumulated in the process of photovoltaic power generation can lead to a reduction in the conversion efficiency of PV modules and severely affect the lifetime of PV modules. In order to investigate the potential of different drenching methods for cooling and improving the efficiency of PV on building rooftops in Guangzhou, a trial bench was set up to test the cooling effect of stand-alone PV modules, and a period of sunny weather with solar radiation exceeding 800W/m^2 was selected for a comparison experiment of stand-alone PV power generation systems. The results show that the cooling effect of several drenching methods is evident, and the use of rocker nozzle drenching has a more potential to improve efficiency, and the growth rate of photoelectric conversion efficiency reaches 9.26% with continuous drenching. Further consideration of the size of the rooftop PV array and the range of the spray nozzles, the rocking arm spray nozzle and the rotating ray spray nozzle are more suitable for practical installation.

Keywords Building rooftop photovoltaic; Photovoltaic module efficiency; Cooling by water; Cooling and efficiency potential

---

高密度街区近零碳排放评估与管控技术研究现状

Research Status of Near-zero Carbon Emission Assessment and Control Technology for High-Density Blocks

傅芷涵，张改景
Fu Zhihan, Zhang Gaijing

傅芷涵，上海市建筑科学研究院有限公司，初级工程师，绿色节能与低碳创新研究院双碳部助理咨询工程师。
张改景，上海市建筑科学研究院有限公司，教授级高级工程师，绿色节能与低碳创新研究院研创中心主任。

摘要 “双碳”战略的深入推进对国内高密度街区的建设提出了更高的碳排要求，本文在国内高密度街区近零碳排放研究面临运行机制复杂、数据获取难、评估技术缺失等挑战的背景下，首先对零碳社区关键绩效指标（KPI）、系统边界以及空间尺度的定义展开了详细介绍，综述了零碳社区全生命周期评估方法，梳理了一系列的评估标准，说明了此类高密度街区近零碳排放的具体定义及评估方法和流程；随后，对国内外现有（近）零碳排街区项目的研究展开案例介绍，梳理了街区项目近零碳研究现状和实现路径。

关键词 建筑环境；高密度街区；全生命周期（LCA）；零碳排放社区

Abstract The in-depth promotion of the carbon peaking and carbon neutrality strategy as put forward higher carbon emission requirements for the construction of high-density blocks in China. Against the background of complex operational mechanisms, difficulties in data acquisition and lack of evaluation techniques in Shanghai' s high-density neighborhoods, this paper firstly introduces the definitions of key performance indicators (KPIs), system boundaries and spatial scales of zero-carbon communities in detail, summarizes the assessment methods for the whole life cycle of zero-carbon communities, compiles a series of assessment standards, and explains the specific definitions of near-zero carbon emissions of such high-density neighborhoods, as well as the assessment methodology and process; then this paper introduces case studies of existing (near-) zero-carbon emission neighborhood projects at home and abroad, and compiles the current status of the research on near-zero carbon emissions of neighborhood projects and the paths of realizing such projects.

Keywords Built environment; High density neighborhoods; LCA; Zero carbon emission communities

基于实际监测的建筑能耗数据管理与应用

Research on Building Energy Consumption Data Management and Application Based on Actual Energy Monitoring Data

孙艾维，林春艳
Sun Aiwei, Lin Chunyan

摘要 建筑能耗数据的统计分析对于建筑用能优化和节能运行管理有重要意义。目前建筑能耗监测系统采集数据质量参差不齐，影响数据分析应用的有效性和合理性，因此建筑运维期能耗数据管理至关重要。本文基于建筑能耗监测平台中数据日常管理实际经验，梳理、分析监测数据准确性的主要影响因素，并通过典型建筑的能耗数据分析案例，讨论平台数据管理优化的要点及数据应用的方法。

关键词 能耗监测；数据准确性；数据管理；数据应用；建筑运行

Abstract Statistical analysis of building energy consumption data is of great significance for building energy optimization and energy-saving operation management. At present, the quality of data collected by building energy consumption monitoring systems is uneven, which affects the effectiveness and rationality of data analysis applications. Therefore, energy consumption data management during building operation and maintenance is very important. Based on the actual experience of data management in the building energy consumption monitoring platform, this paper sorts out and analyzes the main factors affecting the accuracy of monitoring data, and discusses the key points of platform data management optimization and data application methods through the analysis of energy consumption data of typical buildings.

Keywords Energy consumption monitoring; Data accuracy; Data management; Data application; Building operation

孙艾维，上海市建筑科学研究院有限公司，工程师，咨询工程师。
林春艳，上海市建筑科学研究院有限公司，高级工程师，主任助理。

多源空间大数据支持下城镇建筑气候数据的关联网络插值模型研究

With the Support of Multi-source Spatial Big Data Research on Interpolation Model of Correlation Network for Urban Building Climate Data

宋学锟，李红莲，吕文，和玺，霍英博，杨柳
Song Xuekun, Li Honglian, Lü Wen, He Xi, Hong Yingbo, Yang Liu

摘要 本文对国家统计局提供的地级、县级、乡镇级的行政单位与中国气象局所建台站的地理空间位置进行映射分析，从多源的数据产品中挖掘对应的经度、纬度、海拔、城区人口、城区面积、植被指数等空间要素，建立以深度学习方法 LS-GRU 为基础的气象站点 - 乡镇点的关联网络插值模型；选取上海—宝山站、西安—泾河站、乌鲁木齐—乌鲁木齐站在映射关联网络的基础上对其进行插值分析和交叉验证。结果表明：关联网络插值模型的加入大大提高了插值的精度，不同地理环境和不同城镇规模下的待插点与其最佳参考样本台站数量具有空间差异性。

关键词 多源数据；关联网络；深度学习；LS-GRU；最佳参考台站数

Abstract By mapping and analyzing the geographical and spatial positions of the administrative units provided by the National Bureau of Statistics and the stations built by the National Meteorological Administration, the corresponding spatial elements such as longitude, latitude, altitude, urban population, urban area and vegetation index are mined from multi-source data products as a reference, and an interpolation model of the association network between meteorological stations and township points based on the deep learning method LS-GRU is established. Shanghai-Baoshan Station, Xi'an-Jinghe Station and Urumqi-urumqi railway station are selected for interpolation analysis and cross-validation. The results show that the accuracy of interpolation is greatly improved by the addition of the correlation network interpolation model, and the number of points to be inserted in different geographical environments and different town sizes is spatially different from the number of its best reference sample stations.

Keywords Multisource data; Associated network; Deep learning; LS-GRU; Optimal number of reference stations

宋学锟，吕文，和玺，霍英博，西安建筑科技大学，硕士研究生。
李红莲，西安建筑科技大学，教授。
杨柳，西安建筑科技大学，教授。

街区尺度下建筑运行阶段能耗的研究方法与应用

Research Methods and Applications of Energy Consumption in Block Buildings

徐燊，周黄婉瑾
Xu Shen,
Zhou Huangwanjin

徐燊，华中科技大学建筑与城市规划学院，教授。
周黄婉瑾，华中科技大学建筑与城市规划学院，硕士研究生。

摘要 目前对建筑运行阶段能耗的研究多集中在建筑单体尺度，街区尺度能耗是当前亟须研究的方向。本文基于现有学者对街区尺度的研究流程，总结出街区能耗研究方法、物理建模方法、数据分析方法以及应用导控方法四大方面的内容，系统拆解当前对街区能耗研究的途径。本文总结典型学者研究路径，为未来街区尺度研究方法提供参考。

关键词 建筑运行能耗；街区；研究方法

Abstract At present, research on energy consumption during building operation, blocks, and research methods mainly focuses on the individual scale of buildings, and focusing on block scale energy consumption is an urgent research direction. Based on the existing research process of scholars on block scale, this paper summarizes four major aspects of block energy consumption research methods, physical modeling methods, data analysis methods, and application guidance methods, and systematically disassembles the current approaches to block energy consumption research. Summarize the research paths of typical scholars to provide reference for future research methods on block scale.

Keywords Energy consumption; Blocks; Research methods for building operation

---

河南省建筑业碳排放核算与碳达峰预测研究

Research on Carbon Emission Accounting and Carbon Peak Forecasting in Construction Industry of Henan Province

雷馥铭，闫增峰*，倪平安，闫夏，岳英俊
Li Fuming,
Yan Zengfeng,
Ni Ping'an, Yan Xia,
Yue Yingjun

雷馥铭，西安建筑科技大学建筑学院，硕士研究生。
闫增峰，西安建筑科技大学建筑学院，教授。
倪平安，闫夏，岳英俊，西安建筑科技大学建筑学院，博士研究生。

摘要 河南省作为碳排放大省，其中建筑业碳排放总量占据了总排放的近一半，为了按时实现“双碳”目标必须进行有效调控。本文建立河南省建筑业碳排放核算模型，通过多元逐步回归进行影响因素筛选，并改进STIRPAT模型建立建筑业碳排放预测模型，对河南省建筑业碳排放进行核算和预测。结果表明，河南省建筑业碳排放呈稳定上升趋势，并且只有情景模式5可以在2030年达峰，峰值为4.15亿吨CO_2；河南省建筑业减碳重点应放在城市与建筑业发展优化上。本文结果可以为河南省建筑业减碳方向提供理论支撑。

关键词 河南省；建筑业；逐步回归；STIRPAT模型；碳排放

Abstract As a large carbon emission province, Henan Province, in which the total carbon emission from the construction industry occupies nearly half of the total emission, must be effectively regulated in order to achieve the carbon peaking and carbon neutrality goals on time. This paper establishes a carbon emission accounting model for the construction industry in Henan Province, screens the influencing factors through multiple stepwise regression, and improves the STIRPAT model to establish a prediction model for the construction industry, so as to account for and predict the carbon emission of the construction industry in Henan Province. The results show that the carbon emission from the construction industry in Henan Province has a stable upward trend and only Scenario 5 can reach the peak in 2030, with a peak value of 415 million tons of CO_2; the focus of carbon reduction in the construction industry in Henan Province should be placed on optimizing the development of the city and the construction industry. The results of this paper can provide theoretical support for the direction of carbon reduction in the construction industry in Henan Province.

Keywords Henan Province; Construction industry; Stepwise regression; STIRPAT model; Carbon emissions

国内外建筑生命周期影响评价方法比较研究

Comparison Study on Assessment Methodology of Life Cycle Assessment of Buildings of Chinese Standards and International Standards

乔正珺
Qiao Zhengjun

乔正珺，上海市建筑科学研究院有限公司，高级工程师，项目经理。

摘要 本文研究了国内外建筑全生命周期影响分析和建筑碳排放相关标准、导则和指南，对国内外不同标准规范对于建筑全生命周期不同生命周期界定、分析对象、分析边界、分析范围、分析结果的计算要求进行了对比和分析。

关键词 建筑碳排放；全生命周期影响分析；碳排放计算；零碳建筑；低碳建筑

Abstract The paper researches the national and international standards and guidelines for building life-cycle impact assessment and building carbon emissions calculation, studies requirements of building life-cycle impact assessment of national and international green building evaluation standards, compares building LCA methodology, including life cycle stage, LCA scope, LCA boundary and results.

Keywords Building carbon emission; LCA; Carbon emission calculation; Zero carbon building; Low carbon building

---

碳汇导向的城市绿色基础设施规划与量化方法评估

Carbon Sink Oriented Urban Green Infrastructure Planning Methods and Measurement Methods Evaluation

崔博艺，张名凤，吴远翔
Cui Boyi,
Zhang Mingfeng,
Wu Yuanxiang

崔博艺，张名凤，哈尔滨工业大学，硕士研究生。
吴远翔，哈尔滨工业大学，副教授，博士生导师。

摘要 作为复合生态系统耦合作用的综合体，增加城市碳汇具有推动碳中和目标实现的现实意义。基于对森林、农田、河湖水体、湿地、海洋等不同生态系统的碳循环机制和碳库特征进行基础研究与分析，本文总结出城市碳汇机理多样和耦合机制复杂的特性，系统地提出了以碳汇为导向的绿色基础设施优化策略。该研究可为城市绿地绿碳尺度增汇模式选择、土地利用结构调整以及自然资本规划管理提供参考。

关键词 城市碳汇； InVEST 碳储量模块；绿色基础设施规划

Abstract As a complex of coupling effects of complex ecosystem, increasing urban carbon sink has practical significance to promote the realization of carbon neutrality. Based on the basic research and analysis of the carbon cycle mechanism and carbon pool characteristics of different ecosystems such as forest, farmland, river, lake, wetland and ocean, this paper summarizes the characteristics of the diversity of urban carbon sink mechanism and the complex coupling mechanism, and systematically puts forward the carbon sink oriented green infrastructure optimization strategy. This study can provide reference for the mode selection of urban green space scale carbon sink, the adjustment of land use structure and the planning and management of natural capital.

Keywords Urban carbon sink; InVEST carbon storage module; Green infrastructure planning